BATS
IN·QUESTION

BATS
IN·QUESTION

THE SMITHSONIAN ANSWER BOOK

DON E. WILSON

PHOTOGRAPHS BY MERLIN D. TUTTLE

SMITHSONIAN INSTITUTION PRESS
WASHINGTON AND LONDON

Copy editor: Susan M. S. Brown
Production editor: Deborah L. Sanders
Designer: Janice Wheeler

Library of Congress Cataloging-in-Publication Data
Wilson, Don E.
    Bats in question : the Smithsonian answer book / Don E. Wilson ;
  photographs by Merlin D. Tuttle.
        p.   cm.
    Includes bibliographical references and indexes.
  ISBN 1-56098-738-3 (hardcover: alk. paper)
  ISBN 1-56098-739-1 (pbk.: alk paper)
     1. Bats—Miscellanea.   I. Title.
QL737.C5W56   1997
599.4—dc21   96-30035

British Library Cataloguing-in-Publication Data available

Printed in Hong Kong by South China Printing Co.
06 05 04 03 02 01 00 5 4 3

Frontispiece: A spectacled flying fox, *Pteropus conspicillatus*, pollinating black bean, a highly prized hardwood, used especially for veneers.

CONTENTS

.1.
BAT FACTS 1

.2.
BAT EVOLUTION AND DIVERSITY 79

.3.
BATS AND HUMANS 113

Photo gallery appears on pages 74–77.

PREFACE

Bats elicit an immediate and strong reaction from most people. Historically, that reaction has been almost uniformly negative, and bat populations have suffered as a result. In recent years, interest in bats has grown apace with appreciation of them, making bats an excellent topic for this series of books intended to inform the curious public about intriguing groups of animals.

At the same time, there has been a tremendous increase in our knowledge about these animals, especially during the last two decades. Scientists are inexorably attracted to the unknown, and until the 1960s bats represented one of the most poorly known groups of mammals. Their unique attributes—flight and echolocation—added to their allure. Now, early bat biologists have begun to attract excellent graduate students. All of this, combined with the development of new techniques for studying bats, has led to an information explosion about their natural history and a concomitant concern for their conservation.

This book is arranged like others in the series. The first section focuses on the general biology of bats, answering questions dealing with basic bat biology: classification, structure, reproduction, feeding, hibernation, and migration. The second section looks at the evolution of particular types of bats, reviewing the group's enormous diversity. The third deals with bat-human interactions and explores ways people have viewed bats. In addition, the third section contains suggestions on how to pursue your interest in bats should you wish to do so. An appendix provides a complete classification of bats, including scientific and common names for all 925 species and information on the status of their populations. The two-part bibliography includes both general works on bats and specialized scientific references on individual topics.

ACKNOWLEDGMENTS

I wrote this book at home in the evenings over the course of about a year. It was a labor of love, allowing me to describe a wide variety of interesting facts about bats without forcing them into the somewhat turgid prose of a scientific paper. My own love affair with bats began about 30 years ago, when I fell under the influence of Professor James S. Findley of the University of New Mexico. His own keen interest in bats, abiding respect for science, and ability to communicate both have remained a continuing inspiration for me. I would like to say a heartfelt "thanks, Jim," by dedicating this book to him.

Peter Cannell, science acquisitions editor at the Smithsonian Institution Press, was supportive and encouraging at all the right moments. My colleagues Brock Fenton and Tom Kunz suggested additional questions early on in the process. My wife, Kate, read the entire manuscript, catching some remarkable errors and correctly dissuading me from some of my more flowery usages. In addition to providing the superb photographs, Merlin Tuttle read an early draft and offered me valuable suggestions and recent updates on a variety of conservation issues. F. Russell Cole, Adam Potter, and Bernadette Graham contributed considerable time and effort to generating the appendix, and I am grateful for their assistance with this and other projects.

The scientific research departments associated with the world's natural history museums tend to be invisible to the general public, although they provide the underpinnings for all the public exhibits and education programs those museums offer. As it has for everything I have produced for the past 25 years, the Division of Mammals at the Smithsonian's National Museum of Natural History gave me an intellectual home, an unparalleled collection of bats, and superb library resources.

Equally important, my colleagues Alfred L. Gardner and Charles O. Handley, Jr., have shared their extensive knowledge of bats with me on a regular and sustained basis. In return for that support, royalties that accrue from this book will go to the museum for continued support of publication efforts such as this, and to Bat Conservation International to further their efforts on behalf of bats.

INTRODUCTION

Of all mammals, perhaps none are so misunderstood as bats. Because they are nocturnal, secretive, and normally seen only fleetingly as they whip by on a summer evening, bats have taken on an aura of legend. Sometimes this mystery has hampered our ability to appreciate them for what they are.

Although in most people's minds all bats are basically alike, bats actually encompass an incredible variety of forms and lifestyles and are among the most diverse and successful groups of mammals. The 925 species of bats make up almost one quarter of the 4,625 described species of mammals. Bats occur, and are abundant, on all continents except Antarctica. Some bats are efficient predators on insects; others are excellent pollinators of flowers and important dispersers of fruit and seeds. Some eat frogs; others feed on blood. Regardless of these differences, all possess remarkable systems of navigation and flight, unique among mammals. These adaptations allow bats to operate both in the open air and in closed canopy forests, something no other group of mammals has accomplished.

We are coming increasingly to understand the pivotal role bats play in ecosystems all over the world. As more and more of us come in contact with bats, it is vital that we understand the importance of maintaining viable populations of these key species. People are expanding their recreational use of wilderness areas and broadening their awareness about environmental issues in general. Wildlife-oriented activities are becoming more and more important as greater numbers of people are forced by urbanization to live farther and farther removed from natural places and wild things. Development activities continue to take their toll on the natural habitats of all living organisms, bats included. At least 10 species of bats have gone extinct in the past 500 years, and many others have suffered devastating population declines. This list is bound to grow unless the image of bats can be changed so that people recognize their importance and value.

Bats have suffered from bad press for too long. I hope this book will help to counteract some of the negative myths associated with these creatures and allow their more positive attributes to receive the attention they deserve. Information that is accurate and up-to-date can go a long way toward dispelling old stereotypes and allowing a new generation of better-informed citizens to determine the kind of world in which they wish to live. Bats are a significant part of that world, and it is time for us to learn to appreciate them as critical components of our planet's natural resource base. Furthermore, bats are truly amazing creatures, and we should be proud to have them as our neighbors.

.1.

BAT FACTS

WHAT ARE BATS?

Bats, like humans, are mammals (class Mammalia): warm-blooded animals whose bodies are covered with hair and who nourish their young with milk produced by the mothers. Bats share the same senses of hearing, seeing, smelling, and feeling that we enjoy, and they have the added benefit of an exceptional system of navigation and prey detection known as echolocation. Bats range in size from tiny creatures whose forearms measure barely 25 millimeters to Old World flying foxes with wingspans up to 2 meters (Figures 1.1 and 1.2). They bear live young, just as do all other placental mammals. Bats' most remarkable characteristic is flight, a form of locomotion shared with no other mammal. Bats are also amazingly long-lived for

Figure 1.1. (left) A hog-nosed bat, *Craseonycteris thonglongyai*, weighing less than a penny. This is the world's smallest bat; it rests comfortably on two fingers of a human hand. **Figure 1.2. (right) One of the largest bats, Lyle's flying fox, *Pteropus lylei*, being held by a Thai monk.**

such small creatures, with recorded life spans of over 30 years. In addition, they are among the most intelligent of mammals. Indeed, bats are relatively closely related to our own order, Primates.

HOW ARE BATS CLASSIFIED?

Scientists use a system of scientific names called binomial nomenclature, which was developed by a Swedish biologist named Carolus Linnaeus in 1758. Each species of organism has a unique scientific name consisting of two Latin words. The first word in any scientific name refers to the *genus*, and the second word indicates the *species*. Although the generic name can be used alone to describe the group as a whole, the specific epithet is not valid alone; it requires accompaniment by the genus. The scientific name of the species is both those words together.

Genera, in turn, are grouped into categories called *families*. Families are grouped into *orders*, which are grouped into *classes*, which are grouped into *phyla*, which are grouped into *kingdoms*. Each of these categories can also be further divided. Such a hierarchical system allows a single classification to encompass all living things.

This book follows the classification of bats developed by Karl Koopman, a bat specialist from the American Museum of Natural History in New York. There are 925 species, or distinct types, of bats. These are arranged into 177 genera. These genera, in turn, are grouped into 17 families. One family, the Pteropodidae, or Old World fruit bats and flying foxes, composes the entire suborder Megachiroptera. There are 42 genera and 166 species of Megachiroptera, which are distributed throughout the Old World tropics, in Africa, Asia, Australia, and the Pacific islands. The other suborder, Microchiroptera, comprises the remaining 16 families of bats. There are 135 genera and 759 species of Microchiroptera, which are distributed throughout the world (Table 1.1). The appendix contains a complete classification of bats.

As their name suggests, megachiropterans tend to be larger than microchiropterans, but there are additional differences in structure and function between these two groups, which will be discussed in *What Characterizes the Major Groups of Bats?* (in "Bat Evolution and Diversity"). These two suborders constitute the order Chiroptera, or bats, which is one of 26 orders making up the class Mammalia. All members of this class are vertebrates, or animals that possess backbones. Vertebrates also include birds, reptiles, amphibians, sharks, and fish (Table 1.2).

HOW ARE BATS ALIKE?

Although bats are among the most diverse groups of mammals, they are also remarkably uniform in many respects. All bats have the abilities of flight and echolo-

Suborder and Family	Number of Genera	Number of Species
Megachiroptera		
Pteropodidae	42	166
Microchiroptera		
Rhinopomatidae	1	3
Craseonycteridae	1	1
Emballonuridae	13	47
Nycteridae	1	12
Megadermatidae	4	5
Rhinolophidae	10	130
Noctilionidae	1	2
Mormoopidae	2	8
Phyllostomidae	49	141
Natalidae	1	5
Furipteridae	2	2
Thyropteridae	1	2
Myzopodidae	1	1
Vespertilionidae	35	318
Mystacinidae	1	2
Molossidae	12	80

TABLE 1.2. BASIC TAXONOMIC RELATIONSHIP OF BATS TO OTHER ANIMALS

Kingdom Animalia (animals)

PHYLUM CHORDATA (CHORDATES)

Subphylum Vertebrata (vertebrates)

Class Chondrichthyes (cartilaginous fishes)

Class Osteichthyes (bony fishes)

Class Amphibia (frogs and salamanders)

Class Reptilia (snakes, lizards, and relatives)

Class Aves (birds)

Class Mammalia (mammals)

   Order Chiroptera (bats)
   Suborder Megachiroptera
      Family Pteropodidae
   Suborder Microchiroptera
      [All other bat families; see Table 1.1]
   [25 other mammalian orders]

cation, they all have the same general body form, and all have certain similarities in skin and fur, wings, teeth, reproductive systems and patterns, visual acuity, and hearing systems (Figure 1.3).

Form   The basic body plan of a bat is not unlike that of an airplane. The body (fuselage) has a wing attached to either side, a head (cockpit) on the front end, and a tail assembly on the rear. Although they are externally different, the basic internal body plan of a bat is similar to those of other mammals: the head and body are perfectly analogous to those of other mammals, and each wing is supported by the familiar upper arm, forearm, and a set of hand and finger bones (albeit elaborately elongated). The name Chiroptera means "hand-wing" in Latin, and in fact most of the bat's wing is supported by the hand. All five fingers, including the thumb, are represented. Similarly, bats' leg bones are analogous to those of other mammals. However, the upper leg bones, or femurs, have been rotated 180 degrees; thus bats' knee joints bend in the opposite direction (Figure 1.4). This rotation enables bats to hang upside down against a flat surface and still be ready to take flight easily. A membrane stretched between the legs (the uropatagium, or interfemoral membrane) and a tail of variable length (sometimes absent), complete the general body form.

Figure 1.3. A spotted bat, *Euderma maculatum*, in flight from two directions. Note that bats have the same basic structure as other mammals—two arms (wings), two legs, tail, and prominent ears. Their modifications are related to flight.

Skin Their wings and uropatagia are largely naked, so bats' skin is more evident than other mammals'. These membranes are made up of the usual epidermal and dermal layers, closely joined around elastic fibers and an abundant network of blood vessels and nerves. Although somewhat thin and translucent, bats' skin is essentially rubbery, and it is quite durable and resistant to puncture and wear. Small wounds to these membranes heal quickly, with little or no long-term damage. White scars develop as the skin regenerates, but they fade to the normal dark color in a few months. The membranes appear naked to the eye; however, inspection with a microscope shows tiny, transparent hairs on most surfaces. In addition, the skin of the uropatagium is frequently at least partially furred. Hair on the upper surfaces of the membranes probably has to do with airflow during flight, and that on the lower surface of the uropatagium may aid in handling insects captured in flight.

In general, bats' skin is black or dark grayish brown. There are some species, though, with striking black-and-white or black-and-orange patterns on their wing membranes (Figure 1.5). Unlike those of many other mammals, neither the skin nor the fur of bats is used by people. Most bats are just too small to produce a practical

Figure 1.4. A Mexican long-tongued bat, *Choeronycteris mexicana*, roosting in a cave. Note the backward-facing knees, a characteristic of all bats.

amount of fur. However, in some earlier cultures where slave labor was employed—in Inca and Aztec societies, for example—bat fur was used decoratively on clothing.

Skull Bats' heads are extraordinarily variable (see photo gallery, pages 74–77). Although much of this diversity is attributable to surface features, such as nose

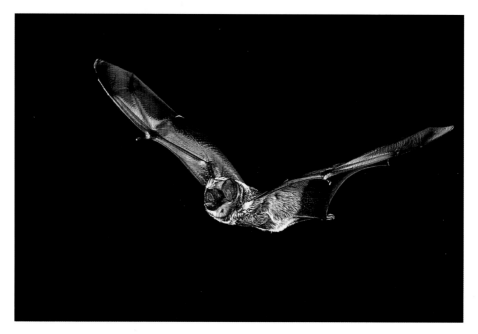

Figure 1.5. A hoary bat, *Lasiurus cinereus*, in flight, showing contrasting color pattern on wings.

leaves, wrinkles, and flaps of skin, the skull itself differs greatly from species to species. This variation forms the basis of much of our system of bat classification. Although the size range of bat skulls is less than that of other orders of mammals, the diversity in shape is remarkable. In general, this variation is the result of a similarly noteworthy diversity in food habits and foraging styles.

The back part of the skull, or braincase, houses and protects the brain and provides attachments for muscles, including some of those used in biting and chewing. Muscle attachment is facilitated by bony structures called crests. A sagittal crest, running along the top of the braincase, and a lambdoidal crest along the back are exceptionally well developed in species with large, strong jaws. Similarly, the zygomatic arches, or cheekbones, range from tiny, fragile bones that may even be absent in some species to enlarged bony plates providing surface areas for muscle attachments.

Jaws   The jaws of bats also reflect their food habits. The most striking variation is in length. Most insectivorous species have moderately sized jaws, with nearly the full complement of teeth (Figure 1.6). Nectar-feeding bats frequently have greatly elongated jaws, with fewer and smaller teeth. This elongation is striking in some species, such as the trumpet-nosed bat (*Musonycteris harrisoni*) and the Mexican long-tongued bat (*Choeronycteris mexicana*), nectar-feeding species found in Mexico (Figure 1.7). Jaws vary in depth as well, presumably also in relation to food

Figure 1.6. (left) A small-footed myotis, *Myotis leibii.* This species has the moderately shaped jaw of a typical insectivorous bat. Figure 1.7. (middle) A Mexican long-tongued bat, *Choeronycteris mexicana.* This species has an elongate rostrum to facilitate its nectar-feeding habits. Figure 1.8. (right) A Leschenault's rousette, *Rousettus leschenaulti,* particularly anxious to get on with its meal. This species' long, thin jaws imply a diet of soft, pulpy fruit.

habits. Long, thin jaws are found in species that feed on soft-bodied insects or soft, pulpy fruit (Figure 1.8).

Other bats, notably some fruit-eating species, have jaws that are considerably shortened and bear curved dental arcades, with the teeth tightly packed on both upper and lower jaws (Figure 1.9). Short, thick jaws are more common in species

Figure 1.9. A great striped-faced bat, *Vampyrodes caraccioli,* perched on a palm frond, eating a ripe fig. These bats carry figs away from the parent tree, eating them in the shelter of palm fronds. By dropping the seeds as they eat, they dramatically increase the number of seedlings that will survive in new locations. Seldom seen and little known, this species roosts in jungle foliage. It is one of several species having bright facial stripes, the significance of which is not yet known.

Figure 1.10. (left) A Jamaican fruit bat, *Artibeus jamaicensis,* feeding on a hard-skinned red almond. In the wild many economically important tropical fruits rely on bats for propagation. **Figure 1.11. (right) A hairy-legged vampire bat, *Diphylla ecaudata.*** One of the three species of vampire bats, this bat has a relatively short, stout jaw useful in administering a slicing wound to its prey.

that specialize on hard-bodied insects such as beetles or fruits that are difficult to penetrate (Figure 1.10). The relatively short, stout jaws of vampire bats are probably related to their need to inflict a slicing wound on their prey to cause the blood to flow sufficiently to the surface (Figure 1.11).

Teeth   Bats have the same types of well-differentiated teeth as other mammals, including incisors, canines, premolars, and molars. Also like those of other mammals, bats' teeth are deciduous; a set of milk teeth is replaced by a permanent set early in the development of a young bat. The milk teeth are highly specialized, consisting of a series of sharply pointed or hooked spicules that allow the young to attach quite firmly to their mothers' nipples. This design enables the youngsters to hold on even during flight (Figure 1.12).

Depending on their evolutionary history and current feeding habits, bats have a variable number of teeth. The original, or primitive, condition for mammals was to have 44 teeth. Small insectivorous bats have 38 teeth, having lost over the course of evolution an upper incisor and an upper and lower premolar on each side. Vampire bats, whose diet of blood requires no chewing, have lost two upper and one lower incisors, three upper and two lower premolars, and two upper and lower molars on each side, leaving a total of only 20 teeth.

Nectar-feeding bats also frequently have reduced dentitions, in terms of both number of teeth and size and complexity of individual teeth. Some species have no lower incisors at all, probably facilitating the extrusion of their long, flexible tongues, used for probing deeply into flowers in search of nectar and pollen.

Figure 1.12. A Brazilian free-tailed bat, *Tadarida brasiliensis*, nursing. The young attach to the nipple using the sharply hooked milk teeth in the front of their jaws.

Variation in bats' dental structure is much greater than in other groups of mammals. This is probably a reflection of bats' great diversity in food habits. As one might expect, carnivorous and insectivorous forms have impressive dental arcades, while those specialized for liquid diets are less strikingly equipped.

Eyes   The great majority of bats, those in the suborder Microchiroptera, have small eyes because echolocation is their major system of orientation. This has led to the common misperception that bats are blind. The expression "blind as a bat" has a certain alliterative appeal, but it is inappropriate. All bats can see, although the extent of their visual acuity is not well known. Most species probably use vision to a certain degree, especially where there is sufficient light. Laboratory studies have shown that many species have considerable ability to recognize and discriminate among patterns and targets of various types.

Megachiropterans are highly visually oriented. They have large eyes, which, combined with expressive faces, make them considerably more appealing to a

Figure 1.13. A Wahlberg's epauletted fruit bat, *Epomophorus wahlbergi,* close up. Note the large eyes and inflated cheek pouches.

public used to the familiar forms of pet dogs and cats (Figure 1.13). Many species of megachiropterans have little or no echolocation ability and thus rely on vision for both navigation and foraging. Their activity periods are frequently crepuscular (around dawn and dusk), as one might expect of visually oriented animals.

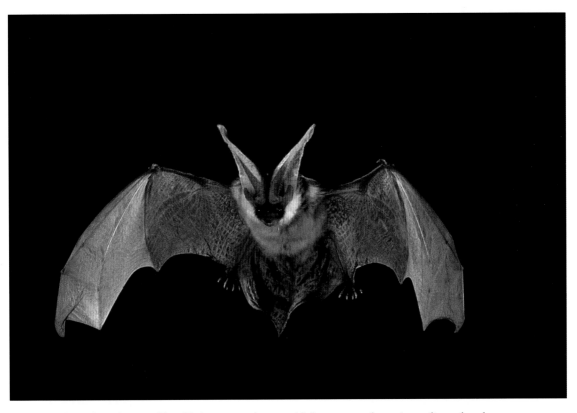

Figure 1.14. A spotted bat, *Euderma maculatum,* **with large external ear pinnae directed to the front during flight.**

Ears Although most mammals use hearing to augment their environmental awareness, bats have developed their auditory systems to an extraordinary extent. Their reliance on echolocation has caused striking evolution in the structure of both their inner and outer ears.

In the visually orienting Megachiroptera, the external ears are of moderate size, rounded or slightly pointed, and without extraordinary additional structures. It is in the echolocating Microchiroptera that ear shape and size are quite diverse.

The major part of the ear is the large external pinna, opening to the front (Figure 1.14). The inside surface is frequently decorated with folds or ridges and may be more or less covered with bands of hair. In addition to the pinna, many species have a small flap at the base of the ear, known as the tragus. It is quite variable in shape and size and forms another useful characteristic for classification and identification of some species. In some species there is also an antitragus, or broad flap continuous with the outer edge of the pinna. Although the exact function of these structures is not well known, it seems likely that the tragus is involved in receiving echoes returning from echolocation pulses.

Ear size varies greatly in microchiropterans (see photo gallery, pages 74–77), from tiny, cup-shaped ears in some families, such as emballonurids and rhinopomatids, through funnel-shaped ears in natalids and thyropterids, to exceptionally long ears in megadermatids and nycterids (see "Bat Evolution and Diversity"). Long ears also tend to be united by a band across the top of the head. In some genera with long ears, such as big-eared bats (*Plecotus*) and spotted bats (*Euderma*), the ears are pleated so they can be pulled back into a curled position while the bats are roosting.

Members of the family Molossidae (free-tailed bats) have ears that are low, joined across the top, and directed forward. This design may be related to their fast-flying lifestyle, because such ears are rigid, streamlined, and seemingly aerodynamic.

Necks Although the neck might seem to be a pretty commonplace feature of most mammals, functioning primarily to form a flexible attachment between the head and body, in bats the neck has been modified in much the same way as the hind limbs. Bat necks, especially in Microchiroptera, are exceptionally flexible and allow the animals to look straight backward while hanging upside down, something bats do routinely while roosting. Remembering that we normally face directly toward our ventral, or front sides, think about the possibility of hanging upside down by your toes, then arching your head back so far that you are staring straight out toward your dorsal, or back side. Special modifications in bats' neck vertebrae allow this marvelous flexibility.

Wings Bats' wings can be thought of as two layers of skin pressed close together, with an inner supporting skeletal framework and the fusiform body causing a bulge in the midline. This structure is readily apparent in a few species known as naked-backed bats (genus *Pteronotus*), in which the wings are joined at the midline of the back rather than on each side.

The wing membranes contain elastic fibers that help hold the wings taut, and thin sheets of muscle help keep the surface smooth to create a proper airfoil. Flight characteristics are partially determined by wing shape (see *How Do Bats Fly?*).

Biologists use two characteristics to describe the shape and load-bearing capacity (or lifting ability) of bat wings. The first, aspect ratio, is calculated by dividing the square of the wingspan by the wing area. This results in relatively high numbers for species with long, narrow wings and low numbers for species with short, broad wings. Long wings are normally found in bats that are capable of fast, sustained flight, such as those in the family Molossidae, which feed on insects caught in midair (Figure 1.15). Short wings are more often found in bats that need to be quite maneuverable and are common in foliage-gleaning species, which need to fly slowly to pick prey from the ground or vegetation (Figure 1.16).

Figure 1.15. The long, narrow wings of a big free-tailed bat, *Nyctinomops macrotis*.

Figure 1.16. A heart-nosed bat, *Cardioderma cor*, approaching its prey, a beetle. Note the short, broad wings. Recent experimentation has shown that this species has hearing so sensitive it locates its prey by listening to the insect's movements. This bat is homing in on sounds made by the insect's footsteps on the sand.

The second characteristic, wing loading, is calculated by dividing the bat's mass, or body weight, by the wing area. This figure, expressed in newtons per square meter, is normally correlated with aspect ratio. Bats with low aspect ratios (short, broad wings), such as nycterids and megadermatids, also usually have low wing loading. These characteristics are in turn correlated with slow, highly maneuverable flight, which may be used to navigate or forage in dense vegetation (see Figure 1.16).

HOW DO BATS DIFFER FROM BIRDS?

Although bats and birds both fly, they are not very closely related. Both are vertebrates, but they belong to different classes; birds form the class Aves, and bats are part of the class Mammalia.

Birds and bats developed flight independently. Birds rely heavily on vision to navigate and to forage for food. This means that their activities are essentially restricted to periods of daylight. Bats, by contrast, primarily use echolocation to navigate and forage, so they can be active at night. Thus, birds and bats can be thought of as ecological equivalents or counterparts; they share many resources, but they use them during different parts of the 24-hour cycle. Anatomically, birds are easily distinguished from bats because they have feathers instead of fur (Figure 1.17). Furthermore, birds have hollow bones, whereas bats have lightweight versions of the normal mammalian, marrow-filled bones.

Figure 1.17. A Lyle's flying fox, *Pteropus lylei*, backlit in flight. Note how the membranous wings of bats contrast with the feathered wings of birds.

The anatomies of birds' and bats' wings are also fundamentally different. On the one hand, the feathers of bird wings are supported by strong (but light) upper arm and forearm bones, with almost vestigial fingers. Bats, on the other hand (so to speak), have greatly elongated finger bones, which form the primary support for their wing membranes.

HOW DO BATS DIFFER FROM OTHER MAMMALS?

The single characteristic that easily separates bats from other mammals is their possession of wings. No other mammalian group has true powered flight, although several kinds can glide for considerable distances. Gliders such as flying squirrels and flying lemurs are inappropriately named; they do not use true flapping flight. Although in some instances they can glide for hundreds of meters, they are capable of only a downward trajectory.

Another feature that separates microchiropteran bats from most other mammals is echolocation, the superbly developed navigational system that allows them to picture the environment through sound much as vision allows us to see it. A variety of anatomical features associated with flight and echolocation help to differentiate bats from other mammals. We will examine some of these in more detail in subsequent sections.

WHAT IS ECHOLOCATION?

Echolocation is the system of high-frequency sounds and their echoes used by most bats to navigate and to locate their food. Bats emit pulses of sound, then analyze information from the returning echoes to "see" the environment around them, thus avoiding obstacles and finding prey items (Figure 1.18). Although bats' echolocation sounds are produced in the larynx, or voice box, much as are the audible sounds of other mammals, they are ultrasonic, which means they are at frequencies beyond our hearing capability (anything over about 15 kilohertz). They are emitted through the mouth in some bats and through the nose in others.

Ultrasonic sounds are also produced by some insects, and a few other kinds of mammals have been shown to use high-frequency sounds for orientation or for finding prey. These include some shrews and tenrecs in the order Insectivora, and whales and dolphins in the order Cetacea. Although these other echolocators produce calls that are structurally and functionally similar to those of bats, there are clear differences. Bat echolocation is much more elaborately evolved. In marine mammals the echolocation system is quite similar to the artificial sonar systems

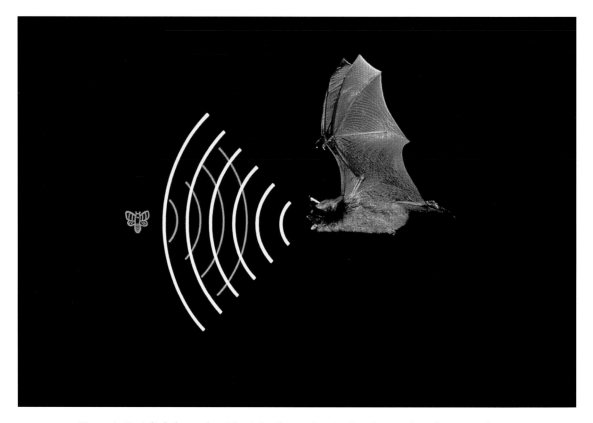

Figure 1.18. A little brown bat, *Myotis lucifugus*, showing how hunting bats detect prey by echolocation. A bat emits high-frequency sounds that rebound back to its ears, enabling it to detect objects as fine as a human hair in total darkness. This bat is such an adept hunter that it may catch up to 600 mosquito-size insects in an hour.

used by submarines. Both use sounds that travel through water to determine distances to other objects.

Echolocation consists of two basic operations: emitting high-frequency sounds, and receiving and analyzing the returning echoes. Sound travels at about 340 meters per second, so by calculating the time that passes between the emission of a signal and the return of its echo, it is possible to obtain information about how far away an object is. Of course, bats do this instinctively, not by deliberate calculation.

Because microchiropteran bats produce echolocation calls in the larynx, these calls can be thought of as vocalizations. By contrast, few megachiropterans echolocate, and those that do click their tongues, so their sounds are not vocalizations. Both suborders also produce audible vocalizations that serve normal communication functions, just as do most mammals.

The returning echoes are received by the ears, and the size and shape of the external ear help amplify these sounds. The echoes vibrate the eardrum, which passes these mechanical vibrations through three small bones in the middle ear to the inner ear, where they are converted to vibrations in the inner ear fluid. These fluid vibrations are in turn converted to electrical impulses, which are carried via the auditory nerve to the brain.

Within the brain the analytical part of the operation includes timing the difference between outgoing and incoming pulses, and comparing a variety of features of both the original pulses and their returning echoes. By assembling and assessing these data, the bat's brain provides information on direction and distance to the target object. The resulting target discrimination is so accurate that some microchiropteran bats are probably able to distinguish among individual species of insects. Although it is cumbersome and time consuming to describe how echolocation works, the process itself is almost instantaneous, with the analysis completed in nanoseconds, in much the same way our brains analyze visual inputs.

In general, echolocation is best suited to relatively short distances. Air absorbs sounds, especially high-frequency sounds, rapidly. Different species of bats use various combinations of frequency and intensity of calls to provide the appropriate combination of target range, or distance, and discrimination, or description. Laboratory studies of the common North American big brown bat, *Eptesicus fuscus*, suggest that it can detect a 19-millimeter sphere at about 5 meters. Other species may be able to detect objects at 10 or 20 meters by using longer calls at lower frequencies.

Curiously enough, bats' exceptional ability to orient without recourse to vision was first demonstrated in a series of elegant experiments performed in 1793 by an Italian scientist named Lazaro Spallanzani. He inserted small tubes into the ear canals of bats so that, by stopping the tubes up, he could block their hearing. He showed that the bats became disoriented and less able to avoid obstacles with a single ear plugged. That bats were producing ultrasonic sounds was not confirmed until about 60 years ago, when a Harvard biology student named Donald Griffin began experiments that opened the modern era of studying this extraordinary ability. Griffin used an extremely sensitive microphone to detect bats' high-frequency calls and to show that the rate of their calls increased as they approached obstacles. Since then the study of echolocation has played a prominent role in our efforts to learn more about bats and how they have adapted to their environments. Scientists now use an electronic apparatus known as a bat detector to distinguish and translate bat calls into audible signals for analysis.

Because humans rely on vision, we think of the world in terms of things we can see. Although it is difficult to conceive of how blind people perceive the world, we know that the images in their brains are not unlike those we produce by vision.

Bats probably produce similarly complete pictures of their surroundings by hearing the echoes from their echolocation calls.

Humans have developed two artificial systems of navigation that are related to the echolocation system of bats. Radar uses electromagnetic energy to analyze distant objects, and sonar uses sound pulses to do the same underwater. Echolocation is similar to radar only in that it is used in the open air. Functionally it is more similar to sonar.

HOW DO BATS FLY?

Although it is fascinating to watch bats as they hang upside down in their roosts, it is on the wing that these animals become truly impressive. If you have ever watched a small bat circling deftly around the trees in your backyard, catching insects on the fly, you must have marveled at its ability to change directions, avoid obstacles, and intercept prey. Flight is the single characteristic that sets bats apart from all other mammals.

Bat flight is controlled by the same forces as all other types of flight, including that of airplanes: lift, thrust, and drag. A bat's wings, especially the parts of the wings closest to the body, form an airfoil. The wind rushing over and under this airfoil travels at different speeds, generating lift that keeps the bat airborne. The sections of the wing farther from the body move both up and down and in a more circular forward and backward motion to generate thrust, the force that propels the bat forward (Figure 1.19). The friction generated by the bat's body moving through the air generates drag, which has to be overcome by both lift and thrust.

Bats vary considerably in their flight characteristics and capabilities. Some species, with high aspect ratios and high wing loading (see the discussion of wings under *How Are Bats Alike?*), fly high, fast, and occasionally quite far (see Figure 1.15). Many such species are migratory, needing to cover large distances in sustained flights of considerable length. Other species, with low aspect ratios and low wing loading, fly slowly, maneuvering through vegetation and other obstacles, and frequently either catch their prey on the wing or pluck it from the surface of the vegetation or the ground (see Figure 1.16).

One characteristic of birds, soaring flight, is uncommon in bats. A major exception, though, is a species of megachiropteran on the small islands of Samoa in the Pacific Ocean. The Samoan flying fox, *Pteropus samoensis*, feeds on a variety of native fruits that it finds by actively surveying its territory each day. That's right—this is the only bat species known to forage almost exclusively in the daytime! Not only is it diurnal but riding thermals and soaring high above the forest canopy have also become part of its routine (Figure 1.20). My colleague John Engbring of the U.S.

Figure 1.19. A red bat, *Lasiurus borealis*, in flight, showing both upstroke and downstroke.

Fish and Wildlife Service and I watched these magnificent animals foraging for several days from an ideal vantage point on a mountaintop on the western Samoan island of Savaii. The bats moved down the mountainside, intermixing gliding and flapping flight as they foraged. They stopped occasionally and slowly but surely worked their way toward the seacoast. To return to the tops of the mountains, the bats rode thermals as hawks do, circling higher and higher until they suddenly turned and used strong flapping flight to head inland at a level just over the highest peaks. Presumably, they conserved considerable amounts of energy by using the thermals to gain the necessary altitude rather than flying directly up the mountainside.

Curiously enough, when we found this same species in Fiji, we learned that it does not soar. There these bats are quite secretive, and although they are diurnal they stick much closer to the forest canopy and rarely fly out in the open. We speculated that this was because of the presence of peregrine falcons on Fiji; these large, efficient predators are missing from Samoa.

Flight is a key characteristic of bats, and it influences all their other anatomical features and many of their physiological functions. This makes bats interesting to a

Figure 1.20. A Samoan flying fox, *Pteropus samoensis*, soaring over the rain forest in Samoa.

variety of scientists. For instance, studies of bats flying in wind tunnels influence the development of new shapes and designs for the flying machines we humans favor.

HOW FAST CAN BATS FLY?

Bats are amazingly adept flyers, but they are not particularly fast compared with the fastest birds. The fastest species, those with high aspect ratios and high wing loading, normally fly high above the vegetation and are not as maneuverable as slower-flying forms. Bats' actual flight speeds in the wild are difficult to measure, but some calculations have been made on bats flying in captivity. Speeds in the range of 30 to 50 kilometers per hour have been reported for a couple of European species, although the measurements were far from precise.

Most North American species measured in free flight under captive conditions were in the range of 8 to 16 kilometers per hour, with the big brown bat, *Eptesicus fuscus*, the fastest at 24 kilometers per hour. Maximum speeds in the wild undoubtedly exceed these indoor measurements, with free-tailed bats in the family Molossidae among the fastest.

DO BATS FLY IN FLOCKS?

Birds are notorious for flying in flocks that can sometimes number in the thousands. In fact, flocking in birds has been the subject of many interesting studies, and several advantages to this behavior have been suggested. In the tropics, foraging flocks of several species move through the canopy, some feeding on fruits and flowers, others on insects scared up by the general ruckus. Flocks also function as early warning systems against predators; the alarm call of one alerts all.

Because of the similarities between bats and birds, it is natural to speculate that bats might also forage in flocks, and considerable circumstantial evidence of such behavior can be mustered. However, providing a true scientific test of this hypothesis has proved more elusive than conclusive. Charles Handley, Al Gardner, and I attacked the problem during our long-term study of the Jamaican fruit bat, *Artibeus jamaicensis*, in Panama.

There we were impressed by the numbers of bats that gathered each night at one or more fruiting fig trees. The bats found these trees with considerably more alacrity than we did, and we wondered if they weren't communicating information about foraging sites either during roosting or by foraging together. Other workers had suggested this type of behavior in flower-feeding bats in Arizona and Brazil.

Our data set included some 9,000 marked individuals of this species, which we captured a total of over 15,000 times. If these bats were foraging in flocks, we should have encountered double recaptures (that is, two bats originally marked at the same site on the same date and recaptured together at another site some interval of time later) more frequently than one would expect on the basis of chance alone.

Actually, even using some rather sophisticated statistical analyses, it is impossible to demonstrate that such pairs occur more frequently than might be expected from chance. Doug Morrison of Rutgers University, who did these analyses, also followed several members of the same harem roost by radiotracking them simultaneously, and he found no evidence that they foraged together.

However, given bats' extremely well-developed social systems, and the fact that we really know very little about the foraging strategies of most species, it still seems likely that some species will eventually be shown to forage in groups.

DO BATS SLEEP?

Not only do bats undergo daily periods of inactivity in the roost but many species sleep the winter away in hibernation. Bats enjoy a type of temperature control, or thermoregulation, known as facultative heterothermy. This means that they can allow their body temperature to decrease if the ambient, or external, temperature drops below their body temperature. Thus, they can effect significant energy

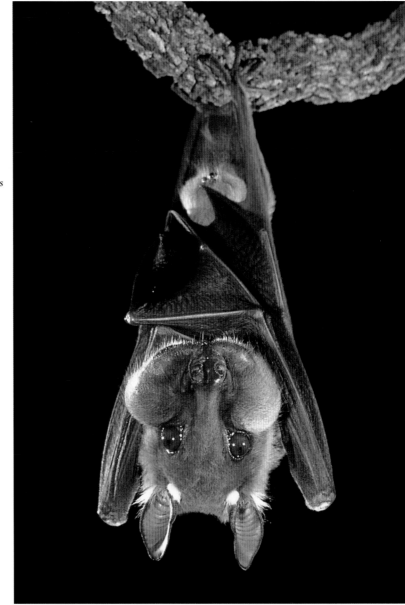

Figure 1.21. A Wahlberg's epauletted fruit bat, *Epomophorus wahlbergi,* **with full pouches.** These large epauletted bats are exclusively African and are extremely important seed dispersers. Only males have pouches, which they fill to help carry their food, but the pouches' primary use is for calling when they are courting females. These bats are not shy and can be seen in the suburbs of Nairobi and other cities and towns, roosting in palm, mango, banana, cypress, and other farm and garden trees.

savings, because mammals expend considerable amounts of energy maintaining constant body temperatures.

In tropical climates bats undergo a relatively normal 24-hour cycle that includes foraging during part or most of the night and a long period of inactivity during the day. Most species also rest for considerable periods during the night between foraging bouts. Whether they actually sleep during these times of inactivity is open to question, but the effect is certainly the same.

ARE ALL BATS BROWN?

Seen flying at night, most bats appear dull. Although the majority are brown or black, there are some spectacular exceptions. Many species have white spots either over the eyes or on the shoulders, where they are known as epaulets (Figure 1.21). In some cases these epaulets are associated with glandular structures, and occasionally such structures are accompanied by bright orange or reddish spots. Countershading—in which an animal is dark on the dorsal, or upper, surface and paler on the ventral, or under, surface—is also common in bats. Presumably this makes them more difficult to see when viewed either from above with a dark background below or from beneath, with a lighter background above (Figure 1.22).

Euderma maculatum, the spotted bat of western North America, is a handsome species with large white spots on a black background, and huge, pale ears that can be folded back when the animal is at rest (Figure 1.23). There are other, completely unrelated, species of black-and-white bats on other continents. Many species of New World fruit-eating bats in the family Phyllostomidae have white lines on their faces, and some have a single white line down the middle of the back (Figure 1.24). Presumably these lines help to break up the shape of the animal's head or body when it is roosting in foliage and might otherwise make an attractive target for a visually hunting predator, such as a snake.

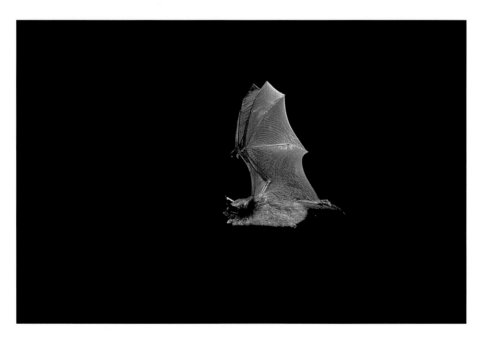

Figure 1.22. A little brown bat, *Myotis lucifugus*, in flight, showing a typical mammalian pattern of countershading: lighter on the bottom and darker on top.

Figure 1.23. A spotted bat, *Euderma maculatum*. This is one of the rarest and most beautiful mammals of North America, ranging from Mexico to western Canada. Its coat of long, silky fur is snow white beneath and jet black above, dramatically accentuated by large white spots on the shoulders and rump. Its translucent pink ears, nearly as long as its body, are the largest of any bat found in the United States, and it has pink wings to match! This elusive animal occurs mostly in arid areas of the southwestern United States, apparently preferring cliff face roosts near water holes. Little is known of its behavior and habits, but it seems to feed mostly on moths, which it captures high above the ground. It is one of the few bats to use echolocation frequencies low enough to be audible to humans. From its scientific discovery in 1891 until 1965, only 35 specimens were known to science. Even now it is one of America's least known animals, but the rarity with which it is observed probably does not reflect its true status in nature. Its habits and choice of roosting sites make the spotted bat unlikely to be harmed by humans.

Figure 1.24. A tent-making bat, *Uroderma bilobatum*, an important seed disperser found in many parts of Latin America. Like most fruit-eating bats, this species carries its meals away from the parent tree, dramatically increasing the number of seedlings that will survive in new locations. Note the distinct white lines on the face, which are accompanied by a single white line down the middle of the back.

Some species of small, insectivorous bats, especially members of the tropical family Emballonuridae, also have white lines on their backs. These species also roost in relatively exposed areas, such as the entrances of caves, overhanging rock shelters, or even between the buttresses of huge forest trees. The wavy white lines on a dark or grayish background probably help to disguise their overall shape. Tufted bats, *Rhynchonycteris naso*, frequently roost on fallen logs along stream banks, and their grizzled gray coloration, with wavy, pale lines mixed in on the dorsum, render them all but invisible until a disturbance causes them to fly (Figure 1.25).

Red bats, *Lasiurus borealis*, are richly colored with a mixture of reds, oranges, and paler tones, and some closely related species are yellow in hue (Figure 1.26). In Africa, the yellow-winged bat, *Lavia frons*, a large megadermatid, has a greenish cast to its fur and yellowish wing membranes, giving it a pastel presence in an otherwise drab family (Figure 1.27). Some species have variegated wings, with dark venation (along the wing bones) and paler colored interstices (between the wing bones), resulting in a striking pattern (see Figure 1.5). *Kerivoula picta*, the painted bat from Southeast Asia, has beautiful wings of black and orange.

Figure 1.25. Tufted bats, *Rhynchonycteris naso,* roosting on a tree trunk overhanging the water of a tropical stream.

A few species of bats are pure white. Some are in the family Emballonuridae, in Central and South America. One of the most attractive is the white fruit bat, *Ectophylla alba,* a tiny frugivore in the family Phyllostomidae, restricted to Central America in the Caribbean lowlands from Honduras to Costa Rica (Figure 1.28). These handsome little animals roost in small groups in tents they create by chewing the midribs of long leaves of banana plants and their relatives, causing them to droop in the form of a tent. The bats are creamy white, with yellowish ears and pink noses. Curiously, the tops of their skulls are protected by a band of melanin, or dark pigment, in the skin under the white fur. This subcutaneous band is unknown in other species of bats but probably helps to protect against solar radiation that may penetrate the leaf forming the roof of their shelter.

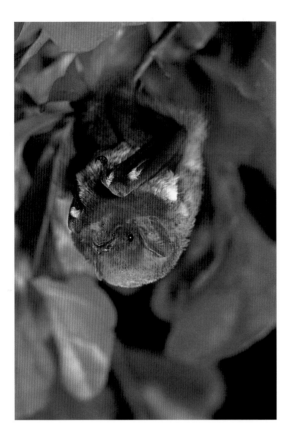

Figure 1.26. A tree-dwelling red bat, *Lasiurus borealis,* curled up in its tail membrane, looking more like a dead leaf than a bat. These bats are found throughout most of the United States, as far north as southern Canada, and south into South America. In the fall they migrate south for the winter. In the summertime they are among the earliest evening fliers and often can be seen around streetlamps, where they hunt moths, among many other insects.

Figure 1.27. A yellow-winged bat, *Lavia frons,* in flight.

Figure 1.28. White fruit bats, *Ectophylla alba*. With long white fur and bright yellow nose leaves and ears, these tiny bats are unique and uncommon. They are found only in the Caribbean lowlands of Central America, typically living in disturbed habitat along the edges of rain forests. White fruit bats roost in groups of 2 to 15 or more, usually under a *Heliconia* leaf, which they make into a "tent" by biting through the leaf's midribs. Inside their day roost the bats are difficult to spot, since the light filtering through the leaf tends to turn the white of their fur to a soft green, hiding them from predators. They do not appear to use the same roost for an extended period of time. As their name implies, white fruit bats are fruit eaters, contributing to tropical forest regeneration.

DO BATS MAKE SOUNDS?

Bats make a variety of sounds, both audible and ultrasonic. The audible vocalizations are used in communication between mothers and their young, among roost mates, and as alarm cries in some species. Hammer-headed fruit bats, *Hypsignathus monstrosus*, in Africa use loud honking calls to attract mates. The males form extended leks, or display grounds, in trees along riverbanks and honk to attract passing females. Jamaican fruit bats, *Artibeus jamaicensis*, emit raucous vocalizations when handled, and those calls attract others, who fly in a swarm around the calling bat, presumably in a reaction that might confuse a predator long enough to allow the bat to escape (Figure 1.29).

As noted, the high-frequency sounds bats produce are absolutely essential to their ability to navigate at night. Echolocation calls vary considerably in frequency and intensity, depending on the species and the function of the calls. Echolocating bats use differences in time and frequency between the pulse emitted and the echo received to discriminate targets such as prey items. They can not only assess the distance to a target but judge details of the target as well. The effort to understand how these sophisticated systems of navigation work forms one of the more exciting areas of bat biology today (see *What Is Echolocation?*).

Figure 1.29. A Jamaican fruit bat, *Artibeus jamaicensis*. These bats produce loud vocalizations when handled. They become coated with pollen as they lap at blossoms such as this balsa flower, and pollen is then carried from flower to flower on the bats' fur. In Latin America balsa wood harvests continue to rely on bats for pollination.

WHAT ARE THE LARGEST AND SMALLEST BATS?

Bats range in size from tiny hog-nosed bats (*Craseonycteris thonglongyai*) that weigh less than a penny to flying foxes with 2-meter wingspans (see Figures 1.1 and 1.2). Both ends of this spectrum are beyond what most people think of when considering bats. If asked to name the smallest mammal, few people would think that it is a species of bat. Similarly, few would guess that bats get as large as the 0.75-kilogram Old World fruit-eating bats known as flying foxes. In reality, bats are on the smallish side compared with many kinds of mammals, such as carnivores and ungulates (hoofed mammals). They are in the size range of small, terrestrial, nocturnal mammals such as shrews and rodents. Also, bats' size distribution is definitely skewed toward the small end. There are lots of species of small and medium-size bats but many fewer really large ones. This distribution is probably influenced by the requirement of being relatively light for flight.

Of the 925 recognized bat species, only about 100 would qualify as really large. Most of these are megachiropterans, but there are a few impressively big microchiropterans as well. In the New World, the largest species is the American false vampire bat, *Vampyrum spectrum*. This species is not a vampire at all but rather a top carnivore, feeding on other bats, birds, amphibians, and reptiles. A related species, the giant woolly bat, *Chrotopterus auritus*, is also carnivorous and nearly as large as *Vampyrum*.

HOW LONG DO BATS LIVE?

Compared with most small mammals, bats are amazingly long-lived. Some individuals may live for 30 or more years, but the average life span is undoubtedly much less than that. The maximum age varies greatly from species to species and is heavily influenced by geography and lifestyle.

Bats that live in the temperate zone and spend more than half of each year in hibernation live the longest. The little brown bat, *Myotis lucifugus*, holds the record for longevity at 33 years. Compare this with most species of rodents, which live for less than one full year.

Longevity records are calculated on the basis of recapture of banded bats. These are normally minimum calculations, because most bats are banded as adults, of indeterminant age. Although banding studies were popular 30 years ago, they are not used as much today because it was discovered that the bands were often detrimental to the bats (Figure 1.30). Bats frequently chew at the bands and may tear their wings or bruise or cut their forearms in the process.

An attempt to avoid this problem led to a different method of banding that was employed successfully in a long-term study of bats at the Smithsonian's Tropical Research Institute in Panama. We used small key-chain necklaces to hold bands where the bats could not reach to chew on them. This method allowed us to calculate survivorship for one of the most common species of tropical fruit bats, *Artibeus jamaicensis*, the Jamaican fruit bat. Their maximum life span has now been recorded up to 9 years. Another tropical species, Linnaeus's short-tailed bat, *Carollia perspicillata*, has been estimated to live upward of 10 years. Similarly, individual flying foxes have lived for over 25 years in zoos. This means that even tropical bats are extremely long-lived compared with other small mammals. Therefore, the theory of hibernation prolonging the life span of bats may be only part of the explanation at best.

WHERE DO BATS LIVE?

As a group, bats are the most widely distributed kind of terrestrial mammal. They are known from every continent except Antarctica. They occur from Alaska in the north to Argentina in the south in the New World, and from near the Arctic Circle to the tip of South Africa in the Old World. They are more diverse in tropical regions but abundant in the temperate zone during the summer.

Figure 1.30. An eastern pipistrelle, *Pipistrellus subflavus,* **hibernating alone.** This bat bears testimony to an ongoing scientific study with its wing band.

Eight families of bats are restricted to the Old World and five to the New World. Only three families are common to both hemispheres; these are Emballonuridae, Molossidae, and Vespertilionidae. Emballonurids are essentially restricted to tropical areas; molossids extend into subtropical regions and partially into the temperate zone; and vespertilionids are widely distributed over all continental landmasses except Antarctica.

Bats are equally eclectic in their utilization of ecosystems. They are abundant in both forests and deserts. Although less common in grasslands and savannas, some species nonetheless occur there. They range in elevation from sea level to 5,000 meters up mountain ranges, although diversity decreases with elevation after about 300 meters or so. They occur in areas with high rainfall and areas with very low precipitation.

One genus of bats, *Myotis*, is more broadly distributed than any other terrestrial mammal genus. Species of *Myotis* occur from Alaska to Labrador and from Norway to eastern Siberia. They range southward to Chile and Argentina in the New World and as widely as Southeast Asia, Australia, and Africa, to the Cape of Good Hope, in the Old World. *Myotis* also represents an ancient, primitive lineage of bats; it is known from Eocene fossils in Europe. This is therefore the most widespread genus of bats in both space and time.

WHERE DO BATS COMMONLY ROOST?

Although bats are extremely widespread on a large scale, they are quite selective in where they live on a smaller scale. Roosts of bats are varied, but most species have fairly specific requirements for where they pass the daylight hours. Bats are also extremely adept at returning to the same roost, not only night after night but year after year in the case of many migratory species. Homing experiments have shown them to be quite capable of returning rapidly and directly from distances of many kilometers. Roosts can be subdivided into day and night roosts, hibernacula (roosts used for hibernation), summer roosts, nursery roosts, feeding roosts, and transient roosts.

Day roosts are the type of bat roost familiar to most people. These are where bats routinely seek shelter during daylight hours, when they are inactive. Caves are probably the most ubiquitous type of day roost used by bats. Most families of bats have at least some species that frequent caves if they are available. Caves provide a variety of advantages, including protection from the sun and predators, and allow bats to conserve energy under relatively constant temperature and moisture conditions.

WHY DO BATS LIVE IN COLONIES?

Although not all bats are colonial, there are some obvious advantages to sharing a roosting site. For many species suitable roosts are in short supply, making colonial life the only possibility. Large colonies offer strength in numbers as well as efficient use of safe havens such as large caves. In addition, roosting in large numbers makes it easier to conserve body heat. Social systems have evolved in conjunction with this habit of colony formation, so many aspects of bats' daily routines are directly related to living in large colonies.

Predator avoidance may have determined the original tendency to form large groups. Certainly it is easier to spot predators when there are more individuals on the watch for them and able to sound the alarm when one is spotted (or heard, as the case may be). At any rate, the habits of living in large colonies and roosting in caves must have developed together. Today some of the most successful species of bats are those that form huge colonies in secure caves.

WHAT IS THE LARGEST BAT COLONY?

Caves also house the largest aggregations of bats known, up to 20 million individuals in the case of some Brazilian free-tailed bats, *Tadarida brasiliensis*, a species found in the southwestern United States (Figure 1.31). Carlsbad Caverns in New Mexico once had 7 to 8 million bats and now houses about 1 million. Bracken

Figure 1.31. An emerging column of Brazilian free-tailed bats, *Tadarida brasiliensis*, one of nature's most remarkable and important animal species. Weighing only about 15 grams each, these bats congregate in the largest colonies of any warm-blooded animal. They migrate south into Mexico and Central America for the winter and return to northern Mexico and the southern United States each spring. In excess of 20 million of these bats roost in Bracken Cave, in the central Texas hill country. Each night from spring to fall, they emerge from the cave in large columns to feed over surrounding towns and farmland. They catch almost 100,000 kilograms of insects—including countless pests—nightly and thus are vital contributors to the health of the local environment.

Cave in central Texas holds about 20 million individuals. Eagle Creek Cave in Arizona is thought to have been the largest bat colony in the first half of this century, with 30 to 50 million individuals. However, their number was reduced to 30,000 in the 1960s as a result of human disturbance.

Bats' propensity to congregate in huge numbers leaves them quite vulnerable to human beings. Cases of vandalism resulting in the death of hundreds of bats were not uncommon just a few years ago. Other colonies were destroyed as a result of superstition or lack of understanding of the importance of bats. In still other cases, innocent exploration of caves at the wrong time of year may disturb hibernating bats, forcing them to use too much energy to make it through the winter.

WHAT IS A BAT GATE?

To combat the problems of bat disturbance and destruction, biologists have developed a system of gating caves using a device that allows the bats to fly through but keeps humans out (Figure 1.32). The design of these gates is quite precise; cave entrances must be blocked carefully because bats have definite preferences about what openings they will use (Figure 1.33). Gates have proved extremely useful in some situations and, when combined with education campaigns to teach the local human population about the necessity of protecting bat colonies, have helped put an end to egregious vandalism and other kinds of bat disturbance.

Figure 1.32. Bat gate construction at Hubbard's Cave, Tennessee.

Figure 1.33. A bat biologist emerging from a protective gate at Canoe Creek Mine, Pennsylvania.

WHAT, BESIDES CAVES, DO BATS USE FOR SHELTER?

Many species of bats use a variety of crevices that provide shelter but are not fully developed caves. Rock crevices are used primarily by insectivorous bats. Similar niches are provided by the loose bark of some kinds of trees and mimicked by fabricated structures such as shingles, shakes, and shutters. Some species of bats routinely squeeze themselves into such tight quarters, which probably provide all the advantages of caves except the constant climatic conditions.

In addition to roosting on and under the bark of trees, many bat species utilize tree hollows as shelters. Such holes provide roost sites in forested areas that might lack caves or rock crevices. Particularly in tropical rain forests, where bats are at their most diverse, caves are few and far between. Most kinds of tropical bats therefore adapt to roosting in hollow trees or directly in foliage.

Many kinds of tropical bats roost in foliage, although such roosts are much less permanent than caves or tree hollows. Foliage roosts range from simple shelters on the undersides of leaves or in dense clumps of foliage to "tents" that are intricately crafted by the bats themselves (Figure 1.34).

One particularly curious adaptation to foliage roosting is manifested by the neotropical disk-winged bats in the family Thyropteridae. These curious little bats have suction cups on their wrists and ankles that allow them to attach firmly to the insides of the leaves of bananalike *Heliconia* plants before the leaves unfurl (Figure 1.35). The bats crawl into the tubes formed by developing leaves and line up in

Figure 1.34. (above) Dwarf fruit bats, *Artibeus phaeotis,* roosting under a *Heliconia* leaf that they have modified into a tent. Figure 1.35. (right) Close-up of the wing disk of a neotropical Spix's disk-winged bat, *Thyroptera tricolor.*

Figure 1.36. A bat biologist monitoring the dawn return of little brown bats, *Myotis lucifugus,* to Canoe Creek Church, Pennsylvania.

small groups in a heads-up posture that allows them to take flight in a hurry if disturbed. This is one of the few species of bats that roost upright.

Several bat species are known to cut leaves to shape them into tents (see Figures 1.28 and 1.34). These bats roost in small groups, and their tents are somewhat ephemeral. A group may use several tents in a small area, switching from one to another after a few days or weeks. The tents provide protection from sunlight and rain and are probably quite secure from most predators as well.

Many kinds of bats have adapted structures created by humans as roosting sites (Figure 1.36). Bats on all continents use attics, and their presence can cause considerable consternation to the occupants of the buildings.

WHY DO BATS HANG UPSIDE DOWN?

The answer to the question why do bats hang upside down involves the same sort of speculation necessary to discuss bats' origins. One plausible scenario is that the habit of hanging upside down developed hand in hand, so to speak, with flight. If bats began as gliding animals, their original takeoff sites were probably under the upper branches of trees, so all they needed to do to become airborne was let go.

Figure 1.37. A lesser horseshoe bat, *Rhinolophus hipposideros*, hibernating while hanging upside down, wrapped in its wings.

This scenario would also suggest that birds developed flight by running and jumping between perches rather than by gliding.

Gliding is basically how most bats launch themselves today. They drop a meter or so before gaining the necessary momentum to propel themselves forward. Although some bats are able to take off from the ground, most are not. Therefore, hanging upside down enables them to escape from potential predators much more quickly and efficiently than they could from an upright posture.

Hanging upside down is also useful in the roost. Hanging from the ceiling of a cave puts bats out of the reach of most predators. With forelimbs modified into wings, a bat's ability to locomote normally is limited. It is certainly beneficial to be able to fold the wings out of the way, or even use them to keep warm, and to rely on the feet to hold the body in place (Figure 1.37).

HOW DO BATS REPRODUCE?

Bats are mammals, so their reproduction is similar to that of all mammals. Fertilization is internal, gestation averages about two months, and the young are born naked and helpless.

Male bats have a conspicuous penis, and the testes move readily from the abdominal cavity to a scrotal position (Figure 1.38). Although females of some species have an external clitoris resembling a penis, which can make sexing them slightly difficult, they also have axillary teats, and in general it is relatively easy to tell male bats from females in the hand.

There is tremendous variation among bats in times of mating, duration of the breeding season, and the relationship of hibernation and migration to ovulation and fertilization. In most species of hibernating bats, copulation occurs in the fall, when swarming flights of males and females are common. Swarming behavior involves increased flight activity, frequently near the hibernaculum, just before entering hibernation. After mating, however, there are various pathways to the production of young the following spring. In some species, such as long-fingered bats (*Miniopterus*) in the Old World, ovulation occurs simultaneously with copulation and fertilization ensues immediately. Development proceeds quite slowly all winter, and the young are born in the spring.

Other species practice sperm storage, whereby sperm that are deposited during copulation in the fall are stored in the female reproductive tract until spring, when ovulation and subsequent fertilization occur. This is the common pattern for many species of Vespertilionids in the temperate zone. In some cases males retain spermatozoa in enlarged epididymides that lie along either side of the tail. They may occasionally copulate during the hibernating period when aroused and may copulate again on emergence from hibernation in the spring. All these mechanisms help ensure that the females become pregnant soon after emergence from hibernation, thus allowing maximum usage of the relatively short period of activity available in temperate zone summers. Mating in spring also allows insemination of females born the previous year, which may not be ready to breed the fall after their birth.

In other species, ovulation and fertilization may occur in the fall but development not begin in earnest until shortly before emergence in the spring. This also would seem to be an adaptation to allow maximizing the time available during the active period in the summer.

In migratory species, mating may be restricted to a short period in the early spring when the bats have reached their summer roosts. The physiological demands of migration may not allow for the added burdens of simultaneous reproductive or developmental activities. In species such as the Brazilian free-tailed bat (*Tadarida brasiliensis*), most births take place together in late spring or early summer.

Figure 1.38. A male straw-colored fruit bat, *Eidolon helvum,* with obvious scrotal testes.

Tropical species, freed from the constraints of hibernation and migration, have more variation in reproductive patterns. Some species have restricted breeding seasons, with only a single young produced per year. This is primarily true for insectivorous species, tied to cyclical insect abundance, which ultimately reflects the alternation of wet and dry seasons.

A more common pattern is seasonal polyestry, two or more cycles of reproduction each year. This occurs in many species of neotropical fruit-eating bats in the family Phyllostomidae. For them copulation happens around the first of the year, young are born a couple of months later, then the females undergo postpartum estrus and become pregnant again right away. This extended period of reproduction generally coincides with the major flowering and fruiting season of tropical plants, which form the major food items of these bats. After the second episode of births, these bats enter a period of reproductive quiescence, normally during the heaviest part of the rainy season.

An interesting addition to this cycle is found in the Jamaican fruit bat, *Artibeus jamaicensis*, a species Charles Handley, Al Gardner, and I studied intensively in Panama for several years. In females of this species, postpartum estrus also occurs after the second birth peak, but the resulting gestation period is lengthened by delayed development, much as in the temperate zone species that pass the winter in hibernation. The embryo develops very slowly at first, normally during the time of heavy rains; then development proceeds apace when the dry season commences, after the first of the year.

In some cases the cycle may be extended, so that individual females theoretically can produce three young during a single long breeding season each year. This occurs in the black myotis, *Myotis nigricans*, a small, insectivorous vespertilionid that is the most widely distributed member of its family in the neotropics. I spent a year studying a colony of about 1,000 of these extraordinary little animals on Barro Colorado Island, a biological station run by the Smithsonian Tropical Research Institute in Panama. That year, 1968–69, provided the first year-round, detailed study of a tropical bat, with data gathered every day. The results yielded insights into the details of this complicated reproductive cycle, although we still know little about what cues the onset of copulation.

Recent studies have found that bats' mating systems are far more complicated and interesting than we had suspected. Early work by Jack Bradbury, a professor at the University of California at San Diego, and his students demonstrated that many species of bats have harem systems, in which a single male controls access to several females. Normally harem males defend roosting territories, where they perform complex rituals of visual, vocal, and olfactory displays to court the females and dissuade other males. In some cases, as in African hammer-headed fruit bats (*Hypsignathus monstrosus*), the males form leks, or display grounds, where they call loudly to attract passing females.

Courtship displays usually involve wing flapping, sometimes vocalizations, and mutual grooming. Sac-winged bats (*Saccopteryx* sp.) have complex displays that involve the male hovering before the females while holding open a special glandular sac in the front of each of its wings. By shaking their wings, the males presumably waft a pheromonal substance toward the females, and they vocalize at the same time. Similarly, epauletted fruit bats (*Epomophorus* sp.) display the wares from their

A.

B.

C.

D.

Figure 1.39. Courtship and mating in the Gambian epauletted fruit bat, *Epomophorus gambianus*. A. Male attracting a female by means of visual attractants (displayed epaulets) as well as glandular attractants. B. A lured female flying toward the displaying male. C. Male and female interacting just before copulation. D. Copulation.

shoulder glands during complicated courtship routines (Figure 1.39). Other species, such as some megachiropterans, roost in large colonies of both sexes, and the males appear to perform only limited displays at best; occasionally they copulate with females that seem to offer little cooperation. Copulation is effected by the male mounting the female from the rear, frequently holding her with his wings and in some cases grasping some of her neck fur in his teeth (see Figure 1.39D).

In all species studied to date, males play no role in rearing young; their involvement in reproduction essentially ends with copulation. However, social behavior is still one of the most poorly understood areas of bat biology, and additional studies may reveal as yet unsuspected interactions.

HOW AND WHERE DO MOTHER BATS GIVE BIRTH?

Females give birth in the roost site, and bat young are born naked and helpless (Figure 1.40). In some species the mother gives birth while hanging upside down; in others she may reach up and hold on with her thumbs, forming a four-point position and using the uropatagium as a basket to help catch the newborn. Neonates are delivered in breech presentation. The feet become free quickly, and the young bat assists in its own birth by grasping its mother's fur with its claws and pulling its upper body free from the birth canal. The newborn moves quickly down its mother's ventral surface, holding on to the hair of her belly until it finds and attaches to an axillary nipple. The mother may aid this process with movements of her wings and feet, all the while holding fast to her perch with one foot or the other (Figure 1.41).

I was fortunate to observe this entire process while studying the black myotis, *Myotis nigricans*, during the field research for my doctoral dissertation. In somewhat ideal circumstances, my study colony was in the attic of a laboratory building on Barro Colorado Island. Similar observations have been made on horseshoe bats in England. Several species of Megachiroptera have been observed giving birth in captivity.

Baby bats are relatively huge at birth. They may weigh up to 40 percent of the weight of the mother. Imagine a 50-kilogram woman having a 20-kilogram baby and you will have some idea of the problems facing a mother bat! The young are born bottom first. This minimizes the danger of getting the wings entangled in the birth canal, although the wings are poorly developed at birth. The thumbs and hind feet are well developed. The young bat clings to the mother immediately, and

Figure 1.40. A newborn fringed myotis, *Myotis thysanodes,* about the size of the tip of a human finger.

Figure 1.41. A newborn Brazilian free-tailed bat, *Tadarida brasiliensis*, with umbilical cord still attached. The mother finally uses her arm to pull on the young, dislodging the placenta. The placenta and umbilical cord remain attached for the first few hours; then they dry up and fall off.

she helps it attach to an axillary nipple right away. The nipples, tucked away in the armpits, allow the young to ride comfortably on its mother's large breast muscle, and her wing surrounds it easily when they are roosting.

Once the young bat attaches to the nipple, dislodging it becomes quite difficult. Mothers prefer to leave their babies behind when they go out to forage, but they may return several times during the night to feed their youngsters, especially when they are very young. Otherwise the young spend their nights hanging in clusters and their days safely tucked into their moms' armpits, attached to nipples.

HOW MANY BABIES DO BATS PRODUCE IN A LITTER?

By far the majority of bats produce only a single young per litter. Most of the exceptions have twins; only a few species have three or four young at a time. Most

Figure 1.42. (left) A hoary bat, *Lasiurus cinereus*, roosting in a spruce tree with twin young.

Figure 1.43. (right) A mother red bat, *Lasiurus borealis*, with her twin young enfolded in her wings as they nurse. Red bats often give birth to twins and sometimes quadruplets. They will begin flying at three to four weeks of age and will be weaned by their fifth or sixth week. During the day each pup holds on to the mother with one foot and on to her perch with the other. When she goes out to feed in the evening, they will be left behind.

bat species have only two functional teats, making larger litters a difficult proposition.

One prominent exception to this rule is the vespertilionid genus *Lasiurus* (Figure 1.42). The several species in this genus regularly have twins or even three, four, or five young at a time. They also have four teats, making it possible to nurse such large litters successfully. These are for the most part migratory species, and they frequently roost alone in the foliage (Figure 1.43), where they leave their young when foraging.

Curiously enough, the common big brown bat of North America, *Eptesicus fuscus*, tends to have a single young in the western part of its range and twins in the east. The significance of this difference remains unknown, although this is one of the best studied bat species in the world.

HOW LONG DO MOTHER BATS SUCKLE THEIR YOUNG?

In general, bats' intrauterine development is much like that of other mammals, with young bats born looking definitely like bats but unable to fly or care for themselves. Their second period of development and growth, outside their mothers' bodies, is characterized by lactation, or nursing of the young by the mothers.

The length of the lactation period varies from species to species, but it is in the range of a month or two for most species that have been studied to date. Evolutionarily speaking, one would expect lactation to last only long enough to ensure adequate physical development so that the young can begin foraging on their own. Until they can fly, the young are relatively helpless, particularly when left alone in the roost. Once they are large enough to fly, weaning and learning to forage both occur rapidly.

HOW FAST DO BATS GROW?

Young bats grow quickly (Figure 1.44). Their mothers carry them for only a short period before leaving them behind in the roost when they go out to forage. Young bats' size places a great burden on their mothers' wing loading. Little brown bats (*Myotis lucifugus*) reach adult size and begin to fly at about three weeks of age. Big brown bats (*Eptesicus fuscus*) take approximately four weeks to reach flying size.

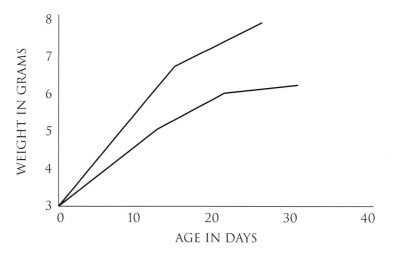

Figure 1.44. Growth rates of gray bats, *Myotis grisescens*, from two colonies. Redrawn from "Population Ecology of the Gray Bat (*Myotis grisescens*): Factors Influencing Early Growth and Development," by M. D. Tuttle, pages 1–24 in *Occasional Papers of the Museum of Natural History, The University of Kansas*, No. 36 (1975).

Once they are capable of flight, the bats quickly become independent of their mothers, although some recent European studies now imply much longer dependence in some species.

Bat youngsters face an interesting set of problems when they begin to fly. They have to learn quickly once they decide to launch themselves, because mistakes can be fatal. Young bats are clumsy at first. Bat biologists studying the development of wing morphology at a colony in Wyoming recently reported that young bats were occasionally seen *walking* back to the roost, appearing considerably disgruntled!

Presumably, young bats initially forage with their mothers, but this is also an area where speculation runs well ahead of actual data. How much training, if any, mothers give their young is still unknown.

HOW DO MOTHER BATS FIND THEIR YOUNG?

One interesting problem faces some mother bats while they are nursing their young. In species that form huge colonies, such as Brazilian free-tailed bats (*Tadarida brasiliensis*), the young are left behind in enormous congregations, called crèches, when the mothers go out to forage (Figure 1.45). Upon returning, a mother bat must find her own young amid the sea of babies on the roof of the cave. Adding to the difficulty is the fact that the babies tend to view all moms as equal and try vigorously to attach themselves to any nipple that comes by.

Until recently scientists speculated that the females "fostered" the young, or simply fed babies at random, rather than relocated their own each time. However, a sophisticated series of experiments using electronic and video equipment coupled with molecular studies of mother-young pairs allowed the bat biologist Gary McCracken and his students from the University of Tennessee to demonstrate that mothers care for young that appear to be their own at least 85 percent of the time, and no tracked mother has ever been seen to switch young. More recent work by the Texas animal rehabilitator Barbara French has shown that mothers who lose their young will help nurse another mother's young.

Mother bats apparently use a combination of cues to find their babies. First, they seem to have some spatial memory of the general area where their young were left. This is not sufficient, however, because the entire mass of young shifts about so that any individual may move a considerable distance during a night.

The females scamper across the babies, fending off youngsters trying to attach to their nipples with quick blows of their wings; all this is done upside down. Mothers use a combination of vocalizations and olfactory cues to locate their babies and quickly allow them to attach to their nipples. Video footage of a mother bat wear-

Figure 1.45. A mother Brazilian free-tailed bat, *Tadarida brasiliensis*, among newborn babies, one of which is hers. Mother and baby will spend up to an hour getting to know each other's scents and vocalizations. Each mother usually produces just one young per year. Amazingly, she finds and nurses her own baby among literally tons of others.

ing a small light-emitting tag and scampering upside down across the hanging heads of thousands of babies was the hit of a recent North American Symposium on Bat Research.

WHAT DO BATS EAT?

As we have noted, bats have a remarkable variety of food habits. It is possible to arrange these feeding types in a hierarchical classification much like the system that we use to classify bats taxonomically. The first major division is between bats that feed on animals and those that feed on plants. The plant-feeding types may be subdivided into those that feed on fruit and those that feed on flowers. Some of these species also occasionally feed on leaves, a behavior about which we still know very little and understand even less.

The animal feeders can be subdivided into those that feed on insects and those that feed on vertebrates. The insect feeders can be further divided into those that

Figure 1.46A. A Gambian epauletted fruit bat, *Epomophorus gambianus*, taking a fig. Most bats prefer to carry fruit away from the tree before eating, apparently to avoid predators. Over several nights bats may carry more than a ton of seeds from a single wild fig tree, dramatically increasing the number of seedlings that will survive in new locations. African epauletted fruit bats, genus *Epomophorus*, are megachiropterans, unrelated to the New World epauletted bats, genus *Sturnira* (see Figure 1.46B), although they are ecological equivalents.

capture flying prey and those that take their prey from the surface of the vegetation or the ground. Vertebrate feeders come in two additional subdivisions, those that feed on whole animals and those that specialize on blood. The whole-animal feeders can be further subdivided into those that feed on terrestrial vertebrates and those that feed on fish.

Fruit Eaters   Fruit eating (frugivory) is a specialized food habit that is found in two similar but taxonomically quite distinct families of bats. The Old World pteropodids (megachiropterans) and the New World phyllostomids (microchiropterans) both have members that are highly specialized for frugivory (Figure 1.46). Species in both groups are critically important to rain forest ecosystems in

Figure 1.46B. A yellow epauletted bat, *Sturnira lilium*, approaching the fruit of *Solanum rugosum*, belonging to the tomato family. These bats are important seed dispersers in the New World tropics. New World epauletted bats appear to rely heavily on this plant for food. *Solanum* shrubs are known as pioneer plants because they are among the first and most abundant to appear on cleared land in tropical forests—whether the clearing is natural or was created by humans. These plants are quick to grow and mature, attracting birds and bats that drop even more seeds, thus accelerating forest regrowth and adding to diversity. Studies in both the Old and New Worlds show that, whereas birds drop most of their seeds around the fruiting tree and in mature undergrowth, bats usually drop their seeds on cleared land, thus beginning the process of renewal. New World epauletted bats are found in tropical areas throughout Latin America.

both hemispheres in that they disperse the seeds of many kinds of plants. These seed-dispersal services have resulted in the coevolution of some species of plants and bats, in which plants develop a syndrome known as chiropterophilly, or bat loving. The characteristics of this syndrome include fruit that is fleshy and sweet but not particularly strong smelling or colorful (Figure 1.47). The fruits often hang from parallel branches, making it easy for bats to get to them. The fruits frequently either are single-seeded with nutritious flesh around the seed, so that the bats drop the seeds after clearing off the flesh, or contain many tiny seeds that pass unharmed through the bats' digestive tracts. These latter seeds are spread thinly throughout the environment as the bats defecate while flying to and from fruit-bearing trees and roosting sites.

Figure 1.47. Examples of fruits that rely mainly on bats for either pollination or seed dispersal.

The commercial value of the products human beings use from bat-dispersed seeds is considerable and still incompletely known. Estimates are in the hundreds of millions of dollars annually. Bats are also extremely important in reforestation of cleared areas. Their tendency to defecate on the wing means that they can quickly sow the necessary seeds back into clearings adjacent to remaining forested areas where they are feeding. In this bats differ from birds, which generally defecate while roosting and thus are less efficient seed dispersers. The numbers of tiny seeds moved by bats on any given night, especially seeds of pioneer plants, or the first to regenerate in clearings, are enormous (Figure 1.48).

Flower Feeders   Bats that feed on nectar and pollen are equally important in providing ecosystem services. Bats are responsible for pollinating an enormous number of plants, particularly in tropical rain forests in all stages of development. The two families discussed in the previous section, Pteropodidae and Phyllostomidae, have developed ecologically equivalent flower feeders just as they have done with fruit eaters (Figure 1.49). These bats have a series of morphological specializations, such as elongated muzzles and long, extrusible tongues covered with fine bristles, that allow them to feed efficiently on a variety of tropical flowers.

Both fruit and flower feeders are essentially limited to tropical ecosystems because of the unavailability of a year-round food source in temperate zones, beset by harsh winters. Even in the tropics, the annual cycle of wet and dry seasons frequently causes these species to switch favored food sources as the seasons progress. In doing so, some may also migrate from one area to another, following the plants' flowering and fruiting cycles.

Plants that rely on bats for pollination have also developed special characteristics to make themselves more appealing to bats. Bat-pollinated flowers have a char-

A.

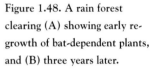

Figure 1.48. A rain forest clearing (A) showing early regrowth of bat-dependent plants, and (B) three years later.

B.

acteristic morphology that includes attributes similar to those seen in bat fruits: pale or no colors, infrequent fragrance, and of course being open at night rather than during the day, unlike many flowers that rely on birds or insects for pollination.

Aerial Insectivores   Bats that catch insects on the fly are the ones most familiar to those of us who live in the temperate zone. These are the whispering shadows we see on summer evenings, zipping through the backyard, dipping back and forth. They are the true masters of echolocation, using this sophisticated sensory system to maneuver around stationary and moving obstacles, to locate themselves spatially, and to navigate between roosts and foraging areas. Furthermore, they use

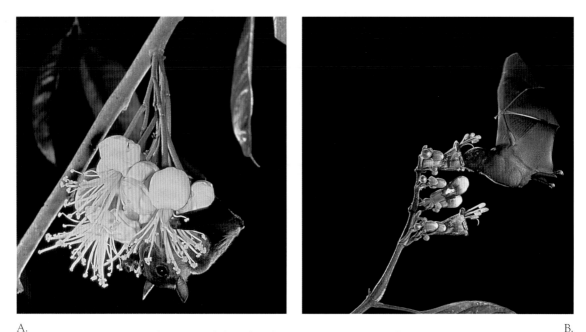

Figure 1.49A. A long-tongued dawn fruit bat, *Eonycteris spelaea,* pollinating a durian flower. The showy white durian flowers exude an unusually heady odor, attracting bat visitors of several species. Flowers open by dusk and begin falling to the ground before midnight. The trees produce a delectable fruit, known by those who have tasted it as the "king of fruit." The great natural historian Alfred Russel Wallace once wrote that it was worth a trip to Southeast Asia just to experience this fruit. Today the market price for a single fruit can be as high as $7 (US) or more in Singapore. Throughout Southeast Asia the durian crop adds as much as $120 million to the economy each year. Yet the bats who ensure that the durian flowers will produce this famous fruit are in decline in many areas and are not protected by local laws. Recent drops in durian production in some areas may be linked to decline of the bats. **Figure 1.49B. A long-tongued nectar bat, *Glossophaga soricina.*** In the New World, bats such as this pollinate flowers of more than 200 genera of tropical plants and are important to the survival of many economically valuable plants.

echolocation to detect tiny insects moving in various directions and are able to calculate intercept distances and angles based on the echoes received from the ultrasonic pulses they emit.

Watching a bat catch an insect is difficult, because all this is done in the dark by animals flying relatively fast, and the action is so quick. All one can really see is the bat making a rapid change of direction, or dip, then continuing on its way. However, high-speed motion-picture photography has allowed scientists to slow this action down, revealing a truly astonishing behavior. Many bats grab the insect in their mouths and simultaneously bring their uropatagia forward with their hind limbs to form a "bug net," into which they push the insect until getting a firmer grip with their mouths. They almost turn a backwards somersault in this process. The

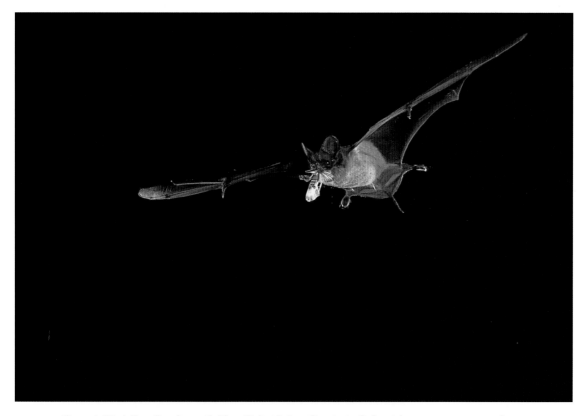

Figure 1.50. A Brazilian free-tailed bat, *Tadarida brasiliensis*, in flight with an insect in its mouth.

action occurs basically within a single wing-beat cycle, and the resultant loss of altitude is negligible.

The economic impact on humans of bats' insect feeding is every bit as significant as that of fruit and flower feeding. Bats consume a staggering quantity of insects every night. A recent Canadian study found that 85 percent of little brown bats (*Myotis lucifugus*) sampled had eaten mosquitoes and concluded that such bats might play an important role in biological control of insect populations. Similar work in Sweden reported mosquito capture rates of over 1,000 per hour for northern bats (*Eptesicus nilssoni*). Calculations for the 20 million Brazilian free-tailed bats (*Tadarida brasiliensis*) in Bracken Cave in Texas suggest that they may consume 225 metric tons of insects in a single night (Figure 1.50).

Such huge colonies of bats leave an interesting "footprint" that can be studied using modern methods of geochronology. Their guano deposits can be cored, dated, and analyzed to yield information on population sizes and food habits over many years. One such colony, in Eagle Creek Cave in Arizona, formerly numbered 30 to 50 million bats that fed mostly over nearby cotton and alfalfa fields, with roughly

Figure 1.51. A Niceforo's big-eared bat, *Micronycteris nicefori*, with a roach in its mouth.

two thirds of their diet consisting of crop pests. Another recent study determined that a colony of 150 big brown bats (*Eptesicus fuscus*) ate enough cucumber beetles over the course of a summer to protect local farmers from some 18 million of their offspring, known as rootworms.

Foliage Gleaners   Less well known are the bats that feed by selecting large insects from the ground or vegetation. These species are found in many families, and they tend to have a set of characteristics related to their foraging habits, including big ears, broad wings, and slow, highly maneuverable flight. They also have well-developed echolocation systems in order to detect their prey against a highly complex background.

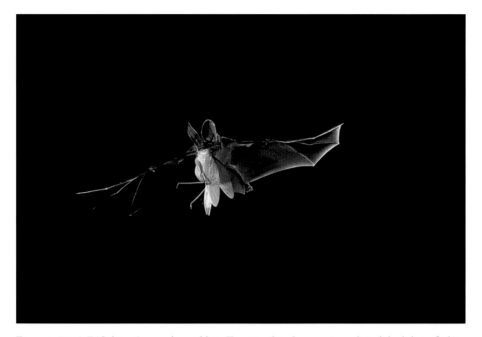

Figure 1.52. A D'Orbigny's round-eared bat, *Tonatia silvicola,* carrying a katydid while in flight in a Panamanian rain forest near the Smithsonian Tropical Research Institute. This species hunts and identifies insects by listening for their mating calls. Recent research has shown that to avoid being caught by hunting bats, some katydid species have developed a system of silent communication through body vibrations called tremulations.

Foliage gleaners typically fly down, grasp their prey off the surface with their large teeth, and move to a nearby feeding roost or temporary perch to eat it. There they frequently eat only the body, leaving the insect wings to fall. Thus the curious investigator can study their food habits by collecting these remains and analyzing them. One such roost on a small island in the Panama Canal allowed me to study the food habits of the hairy big-eared bat, *Micronycteris hirsuta,* at a time when we knew virtually nothing of its food habits (Figure 1.51). The remains, which I collected by putting a sheet down and returning regularly to gather its contents, also confirmed that this species switched to occasional fruits in the dry season, when insects are scarce and fruit is abundant.

Some of these gleaning species can also detect the sounds made by the prey itself. Many species of insects generate unique sounds, and almost all make some slight noise just by moving through the environment. These sounds may be sufficient for a highly efficient, auditory-oriented predator such as a foliage-gleaning bat to detect against the general background noise (Figure 1.52).

Foliage gleaners do not restrict themselves to insects. Some species, such as the

Figure 1.53. A pallid bat, *Antrozous pallidus*, eating a scorpion. This bat is unique for its habit of catching scorpions, unharmed by their stings. Bats have developed varied food preferences, allowing several species to live in the same area with minimal competition for food. A few insectivorous bats, such as this one, forage for prey on the ground. Others glean insects from foliage, but most capture aerial prey.

pallid bat (*Antrozous pallidus*) of the southwestern United States, capture other arthropods, such as centipedes and scorpions, from the surface of the ground (Figure 1.53). This type of food habit may be a secondary specialization in bats' evolutionary history. Bats probably evolved from insectivorous ancestors, and the development of powered flight allowed them to exploit a rich source of food in flying insects. Did the foliage gleaners continue to feed on crawling insects and simply adapt their new powers of flight to better methods of capturing them? Or is foliage gleaning an offshoot of aerial insectivory that allows some species to use significant packets of energy in the form of large, slow-moving insects taken directly from the ground or vegetation? The morphological specializations are probably the same in either scenario, making speculation difficult. It is evolutionary puzzles such as this that keep bat biologists studying the seemingly endless details of bat morphology, behavior, ecology, and physiology.

Carnivores   Bats that feed on other vertebrates may be considered the top carnivores of the bat world. Species such as the American false vampire bat, *Vampyrum spectrum*, with large bodies, broad wingspans, and sophisticated dentition, are

Figure 1.54. A frog-eating fringe-lipped bat, *Trachops cirrhosus*. These bats can distinguish between poisonous and edible frogs, and also locate frogs by their calls. The frogs must call to attract mates, but they face the possibility of attracting a fringe-lipped bat instead. Recent studies demonstrate the major influence of bat predation on frog courtship behavior.

among the most impressive bats. These animals are not common anywhere, but they are widespread in tropical rain forests from Mexico to Brazil. They seem to occur in pairs, and probably an individual or pair needs a certain amount of foraging territory in order to survive.

These bats are attracted to other bats and may occasionally be captured in mist nets because they come to investigate bats that already have been captured. One night in Costa Rica, I sat on the forest floor very near a net with a cloth bag containing live bats hanging from my belt. An adult male *Vampyrum spectrum* flew so low toward the bag that it struck the bottom of the net and was momentarily thrown to the ground right next to my leg. I am sure that it was drawn to the area by the cries of the bats in the bag and was intent on making a meal of them.

One particularly interesting vertebrate predator is *Trachops cirrhosus*, the fringe-lipped bat of the neotropics (Figure 1.54). Studies by Merlin Tuttle of Bat Conservation International have shown that this species is a highly efficient predator on tropical frogs. These bats locate breeding accumulations of frogs by their mating calls and pluck them from the surface of the water or surrounding land by flying in low and grabbing them with their well-developed canines and other teeth. The bats are so efficient that they can distinguish edible from distasteful frogs without even completely grasping the latter.

Many species of bats around the world have developed the habit of feeding on other vertebrates. They use a wide variety of prey, including birds, lizards, frogs, and other bats. Members of the family Megadermatidae are efficient predators in Asia and Africa. In Latin America, several members of the subfamily Phyllostominae fill this niche.

Fish Eaters   Several species of bats are specialized to feed on fish. These are found on almost all continents, and they represent several families. This diversity suggests once again that such behavior has arisen independently in disparate groups, another example of ecological equivalence. Fish eating is such a curious specialization that these independent evolutionary pathways must have followed similar intermediate stages.

If we envision bats that feed on flying insects emerging from underwater hatches, it is easy to imagine large numbers of bats feeding quite close to the water surface. Indeed, this type of feeding behavior is common in several species of aerial insectivores today. It is not a great stretch to think of these animals actually taking insects from the surface of the water, and in fact some species of bats do so today. One species in the neotropics, the long-legged bat, *Macrophyllum macrophyllum*, seems particularly well adapted to this niche.

From this stage it might be a logical step to begin taking small fish from the water surface as well, and perhaps this is how the fish-eating species evolved their strange behavior. Fishing bats are marvels of echolocation and coordination. They use echolocation to detect minute ripples on the water surface caused by small fish feeding just below or at the surface; then they swoop down, gaff the fish with the specially enlarged claws on their hind feet, and fly away with them to a feeding roost. This habit is particularly well developed in the greater bulldog bat, *Noctilio leporinus* (Figure 1.55).

Blood Feeders   Surely the most bizarre feeding habit of any vertebrate is that of the three species of vampire bats (see Figure 1.11). This group, the subfamily Desmodontinae, is restricted to the New World tropics. Vampire bats are not particularly large, but they have a variety of morphological specializations related to their curious feeding habits.

These bats are particularly adept at terrestrial locomotion and can walk, hop, jump, and scramble about on the ground as competently as many four-legged animals. They have elongated thumbs, which aid considerably in their ability to move

Figure 1.55. A greater bulldog bat, *Noctilio leporinus*. Found in Latin America, these bats are members of the family Noctilionidae, also called fishing bats because of their food habits. These relatively large bats fly over freshwater pools and slow-flowing rivers, as well as saltwater lagoons, in search of prey. They use their echolocation to detect ripples made by surface-swimming fish. The bats gaff fish up to 8 centimeters long out of the water with their disproportionately large feet. The fish are either eaten in flight or stuffed into cheek pouches and eaten later at a feeding roost.

about. They are capable of jumping straight up and taking flight from the ground surface, a trait uncommon among other bats.

Vampire bats' dentition is highly specialized as well. Their incisors are canted forward to a great degree and are razor sharp. The bats use these teeth to make a small nick in the prey, and an anticoagulant in the bats' saliva allows the blood to flow. The canines are also well developed, but the remainder of the teeth are reduced or absent. The tongue is short and thick, with a pair of lateral grooves running the length of it. The tongue is run quickly in and out of the wound, causing the blood to flow by capillary action up the grooves and into the bat's throat. Thus, vampire bats are more properly described as blood lappers than as blood suckers.

Curiously enough, these animals, which have been associated with horror stories for so long, have come full circle in their utility to human beings. A drug company has recently patented a new blood thinner called Draculin, developed from research on the natural anticoagulants in vampire bat saliva. This also provides an excellent example of the ongoing necessity of basic research, although the utility of such studies may not be obvious from the beginning.

Although they are unknown outside the New World tropics, vampire bats were once much more diverse than they are today. We know of fossil species that were also considerably more widespread in past times when the climate was milder and they extended well into North America. Some of these species were also somewhat larger than the three species known today. Two of the three modern species are much less common than the third, and both are specialized to feed on the blood of birds.

The common vampire bat, *Desmodus rotundus*, is widespread from northern Mexico to Argentina and sometimes abundant in cattle-growing regions. In undisturbed habitats, these bats are quite uncommon, and they probably originally had to rely on the relatively few species of large mammals found in tropical rain forests. The development of modern agriculture and the abundance of livestock have greatly increased the populations of common vampires, to the point where they are considered a major pest in many areas. This is a case in which the activities of *Homo sapiens* have actually caused the population of a species to increase.

The other two species of vampire bats, the white-winged vampire, *Diaemus youngi*, and the hairy-legged vampire, *Diphylla ecaudata*, are much less common. Both tend to specialize on bird blood and so pose much less of a threat to humans. However, in some areas they do prey on domestic fowl such as chickens.

The vampire legends of Eastern Europe have nothing to do with vampire bats of the New World. The legend of vampires returning from the dead to feed on their victims' blood was well developed before the discovery of vampire bats in the New World. Early naturalists knew that there were bats that fed on blood because these animals were well known to the indigenous peoples in their area. These bats do occasionally feed on people, although we are not their favorite prey. Occasionally, someone sleeping in the open or in an unscreened room may awaken to find a small wound that continues to seep blood for an inordinately long time. Strangely enough, such people are often fed upon more than once, while others sleeping in the same room may be ignored. Probably the bats remember successful prey items and return to them regularly. If a sleeping person does not awaken during a feeding bout, he or she will probably be visited again.

Early naturalists often were unable to distinguish which animals were the vampires, and they sent back to the Old World specimens of many kinds of bats with notes indicating they were probably vampires. Hence a variety of genera of New World bats have names that suggest they are vampires: *Vampyrum, Vampyrodes, Vampyressa*. Actually, none of these are vampires. The generic names of true vampires are not so suggestive: *Desmodus, Diaemus, Diphylla*. (See the discussion of vampire bats in "Bat Evolution and Diversity" for more information on these.)

Figure 1.56. A Linnaeus's short-tailed bat, *Carollia perspicillata,* (A) approaching the fruit of *Piper hispidum,* (B) plucking it from the plant, and (C) flying to a nearby feeding roost. This bat may eat up to 35 of these fruits in a night, ingesting and dispersing thousands of seeds.

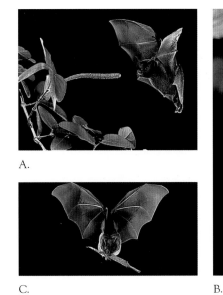

A.

C.

B.

HOW DO BATS FIND FOOD?

Like other animals, bats are faced with the problem of finding enough food to survive to the following day. They have to be aware of their environment to the extent that they have a reasonable idea of where to search for food. They also have to be able to locate specific food items against the general background of complicated environmental stimuli. Finally, they have to catch, subdue in some cases, and eat the individual food items.

The search for food depends on the food habits of the species in question. Fruit-eating bats may search generally through the surrounding environment for fruiting trees of a suitable species. *Carollia perspicillata,* Linnaeus's short-tailed bat, forages by flying through suitable habitat searching for its favored food, the fruits of *Piper* plants, which are scattered more or less randomly through the undergrowth in tropical regions (Figure 1.56). The fruits of most species of *Piper* are upright on the branches, making them readily available to the bats, which oblige the plants by dispersing their seeds widely. Commercial black pepper is harvested from another species of this genus.

In Panama, where we studied the Jamaican fruit bat, *Artibeus jamaicensis,* a different strategy is used. These bats primarily target fruiting fig trees, and they know the locations of suitable trees in their home range. They probably take note of surrounding trees each night on their foraging flights and normally fly directly to a tree with fruits at the appropriate stage of ripening.

Insectivorous species have options as well. Many species feed on flying insects, and they normally fly directly to suitable habitat patches, such as over water or

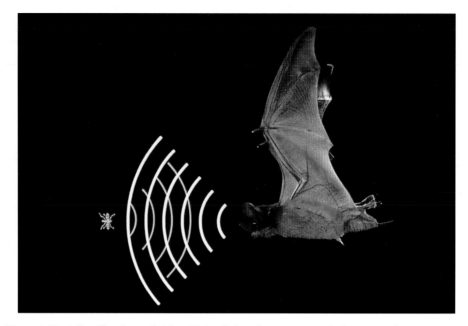

Figure 1.57. A Brazilian free-tailed bat, *Tadarida brasiliensis,* using echolocation to home in on a flying insect.

around light sources, where their prey may congregate. Foliage gleaners have to know how to locate suitable patches of vegetation or ground surface in order to hunt for specific prey items. Still others hang from an appropriate perch and wait for prey items to come by. They then sally forth to effect the capture, much as do many species of insectivorous birds.

Specialists, such as those that feed on other vertebrates or fish, or the blood-feeding vampires, also have to locate proper habitat patches in order to begin the detection activities that allow them to find individual items. Vampire bats in natural forest areas probably have some difficulty in locating suitable large prey, because large wild mammals such as tapirs are becoming more scarce. However, in agricultural areas vampires fly directly to cow pastures or corrals containing domestic livestock.

Bats that feed on insects and other animals almost all use echolocation to detect individual prey items (Figure 1.57). Bats that prey on flying insects are able to detect differences in the echoes returned from the insects' wings. Not only can they locate the insects but they can determine from the echoes whether an insect is flying toward them or away from them and are able to achieve the proper intercept pathway to capture the insect. Some species, particularly in the Old World, are so specialized that they key in on insects' wing fluttering. Those that feed on insects on vegetation or on the ground have to be able to detect the prey item against the background, and their ability to differentiate the echoes returning from these items allows them to accomplish this task. They also rely on sounds from the insects themselves, including footsteps, mating calls, or even chewing sounds.

One specialized predator detects prey by listening for the calls made by the prey items themselves. *Trachops cirrhosus*, the neotropical fringe-lipped bat, locates ponds of calling frogs by listening to the sounds male frogs make to lure females. The bats have keyed in on the calls of some species of frogs sufficiently to have caused changes in their calling pattern, as the frogs attempt to escape predation by the highly efficient hunting bats.

Scientists actually know less about how bats detect ripe fruit or nectar-filled flowers than we do about animalivorous forms. The fruit and flower feeders probably use echolocation to locate items generally, but they likely rely on olfactory and chemical cues to determine whether an item is suitable to eat.

HOW SMART ARE BATS?

Intelligence is a relative trait and one that is quite difficult to compare across groups of animals. There are no direct measures of intelligence in animals, but various ways of estimating it have been proposed. By any measure, mammals are thought to be "smarter" than other kinds of animals, and bats are among the most intelligent mammals.

The most common way of approaching intelligence in animals is to measure brain size in comparison to body size. The idea is that bigger brains connote higher intelligence, or at least more potential "brainpower." Brain size plotted against body size results in a highly significant positive correlation, meaning that larger animals have larger brains. This is to be expected, but it is in the comparison of such lines that more interesting details appear.

In plotting such comparisons, we find that fruit-eating bats routinely have larger brains per body weight than do insectivorous ones. This might seem surprising at first, because well-developed brains surely must be required to echolocate well enough to catch insects on the wing. And why should it require even more brainpower to find ripe fruit and flowers?

Actually, although echolocation has been honed to an elegant precision by millions of years of evolution, the process of detecting and discriminating prey is essentially quite stereotyped, or automatic, once engaged. By contrast, surveying the environment to find and store in memory the locations of trees bearing ripe fruit, or fruit that will become ripe shortly, is probably considerably more difficult.

Species that feed by foliage gleaning, those with specialized diets such as fish or blood, and those that feed on other vertebrates also have relatively large ratios of brain to body weight. In a series of studies that involved these types of comparisons,

John Eisenberg and I concluded that large brains are related to a foraging strategy that involves locating extensive patches of high-energy food, which are difficult to predict in time and space.

It is possible to refine these studies to determine the regions of the brain that are particularly well developed. As one might expect, bats tend to be quite well developed in the auditory regions. This is particularly true of the Microchiroptera, for which echolocation is paramount, whereas Megachiroptera have better developed olfactory and visual areas.

Another component to the brain size issue is almost surely related to habitat complexity. Those species that deal with spatially complex habitats, such as rain forests, probably need more and better abilities to store and use sensory information than do those that deal with spatially simple habitats, such as grasslands. This idea has not been well tested in bats, although there are some intriguing hints from studies on other kinds of mammals.

WHAT DO BATS DO IN WINTER?

In the temperate zone, bats are active only during the warm months, when insect populations are high. In the winter they have to either migrate south to warmer climes, as many birds do, or hibernate in cold caves, where they sleep away the winter in a state of suspended animation. During hibernation body processes slow drastically, and energy consumption is minimized (Figure 1.58). Nevertheless, most hibernating individuals arouse one or more times during the winter, which speeds depletion of their energy reserves.

Migratory species may fly thousands of kilometers in search of favorable climates that allow year-round activity (Figure 1.59). The annual trip north is worth the effort because of the superabundance of insects in temperate zone summers. This food supply allows the bats to satisfy the additional nutritional requirements of reproduction and provides a plethora of prey items for the young of the year as they learn to forage.

WHY DO SOME BATS HIBERNATE?

Tropical bats are active year-round. This is possible because they have an ample food supply throughout the year. In the temperate zones that food supply is available for only part of the year. In fact, in the temperate zones the only sufficient food supply for bats consists of the large populations of insects present during the summer.

In the autumn, when insects become scarce because of falling temperatures, bats are unable to find enough to eat. Many species solve the problem by going to sleep

Figure 1.58. A hibernating eastern pipistrelle, *Pipistrellus subflavus*, with its fur covered by moisture droplets. The condensation indicates that the bat's body temperature has dropped to that of the surrounding cave. This species hibernates individually and is the one most often seen by cave explorers throughout eastern North America.

Figure 1.59. An emergence of Brazilian free-tailed bats, *Tadarida brasiliensis*. These bats migrate annually between large nursery roosts in the United States and wintering grounds in Mexico.

Figure 1.60. Endangered Indiana bats, *Myotis sodalis,* clustering close together in a cold cave to maintain proper body temperature with minimum energy expenditure. Of all available caves, only a few are suitable for bats' hibernation. Disturbance of their hibernating roosts can cause these bats to rouse, costing them from 10 to 30 days' supply of stored fat reserves, which must last them through four to six months of no feeding. Because these bats have to congregate in great numbers in relatively few caves, they are highly vulnerable to destruction. Indiana bats have suffered rapid population declines and are in danger of becoming extinct, mostly because of human disturbance.

until spring, when insect populations begin to grow again. Actually, hibernation is a bit more complicated than simply going to sleep for the winter, but the bats do become torpid for long periods of time, during which they allow their body temperature to fall to or near the temperature of their surroundings, as long as that ambient temperature remains above freezing (Figure 1.60).

Most hibernating bats spend the summer months feeding heavily on the abundant insect supply, building up enough fat to provide an energy source during the long winter of torpor. They may fly several hundred kilometers in search of just the right cave in which to hibernate. These hibernacula are used year after year by the same bats. An individual may awaken several times during the hibernating period and even drink water or feed if food is available. However, most species must survive the winter solely on stored fat. They may also undergo occasional bouts of reproductive activity during times of temporary arousal from hibernation (Figure 1.61).

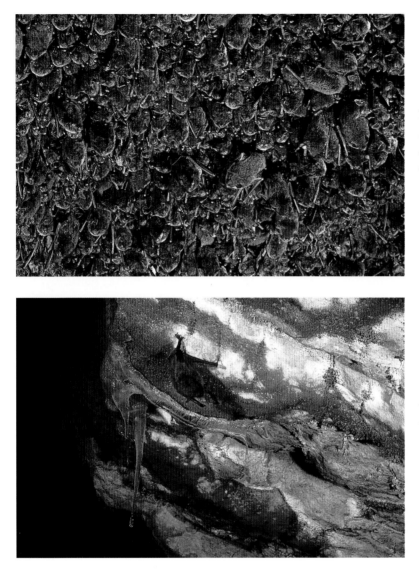

Figure 1.61. A huge cluster of southeastern myotis, *Myotis austroriparius,* hibernating in a cool cave.

Figure 1.62. A solitary big brown bat, *Eptesicus fuscus,* hibernating among the icicles in a cold cave.

WHAT DO BATS DO WHEN THEY HIBERNATE?

Bats do precious little when they hibernate. The bat settles into a comfortable posture, frequently tightly bundled with several of its roost mates. Then it slowly allows its body temperature to fall until it reaches the temperature of the surrounding air. During this process, all the animal's vital functions—pulse rate, breathing rate, and metabolic rate—slow down, and the bat essentially enters a state of suspended animation (Figure 1.62).

Bats are known as facultative heterotherms, which simply means that they allow their body temperature to fall but are sufficiently in control that they can rewarm themselves if necessary. Obligate heterotherms are forced to live with a body temperature the same as the ambient temperature and have to use external mechanisms such as the sun to warm themselves. That is the case with reptiles, for instance, and explains why you see snakes and lizards sunning themselves but will never catch a bat doing so.

WHY DO SOME BATS MIGRATE?

Another characteristic that bats share with birds is migration. Many of the most common bats in the north temperate zone are there only during the summertime. They move northward across North America in the spring, and some species are widespread in Canada during the summer. Then in the fall they move back southward to Mexico, Central America, and perhaps even South America. These bats migrate for the same reason that other animals hibernate: lack of food in the winter. Long-nosed bats in the genus *Leptonycteris* are a good example, moving northward into the southwestern United States in the summertime, following the flowering season of one of their favored foods, agave plants (Figure 1.63).

We know little about these long-distance migratory movements of our common bats. Although thousands of bats have been banded at summer roosts in the United States, very few are ever recovered outside this country, at least partly because less collecting is done in other countries. The Brazilian free-tailed bat, *Tadarida brasiliensis*, forms huge summer colonies of up to 20 million animals in the southwestern United States in the summer, but their winter homes are much less well known. We do know that migration is difficult for them, as experimental work on the effects of pesticides has shown. They tend to accumulate pesticide residues in their body fat as a result of feeding on insects that have been exposed to insecticide-treated fields. Then, when the fat is burned during migration, the residues are released into the bloodstream and may reach the brain and other vital organs, causing debilitation or even death.

One common bat throughout North America during the summer is the silver-haired bat, *Lasionycteris noctivagans*. This species moves through the southern United States in spring and fall, obviously migrating, and its members are easily captured in mist nets over a broad area of the continent during the summer. However, in the winter these large numbers seemingly disappear. Although individuals routinely hibernate in deep cliff-face crevices, in old stumps and logs, and in the cavities of living trees in the southeastern United States, there are no known large concentrations of them as there are of other hibernating species.

Until a few years ago, there were no records of this species from south of the U.S. border. Even now, the species is virtually unknown from Mexico, making long-distance migrations seem unlikely for it. It seems amazing that we know so little about

Figure 1.63. Two southern long-nosed bats, *Leptonycteris curasoae*, approaching an agave flower. This plant relies so heavily on bats for pollination that without them its seed set drops to 1/3,000 of normal.

such a common animal, but following individuals on a migratory flight would be a very expensive proposition. So far funding to do such studies has not been available.

Bats also undertake somewhat simpler and more local migrations. Although many species move dozens or even hundreds of kilometers in search of the proper cold caves in which to hibernate, others may move up and down in altitude as seasonal changes in food supply occur. Conservation programs now take such migratory pathways into consideration, and efforts are made to preserve altitudinal corridors that will allow animals to exist throughout the year without having to leave the area in search of food.

DO BATS FEEL PAIN?

Photographs and descriptions of pallid bats (*Antrozous pallidus*) feeding on scorpions immediately bring to mind the question of whether bats feel pain (see Figure 1.53). The bats do not appear to handle the scorpions in ways that avoid being stung—they crunch them, seemingly oblivious to the stinger that we, as pain-sensitive *Homo sapiens*, dread. Because we feel pain, and because other mammals have nervous systems similar to ours, we assume that they also feel pain. However, there is as yet no way to determine if this is true. Bats' reactions to predators and their efforts to escape when handled all would lead one to believe that they are sensing something that causes them to attempt to flee. But like many other areas of bat biology, this question will have to await future studies for a definitive answer. However, for all practical purposes it seems safe to assume that bats do feel pain and that pallid bats have adapted in a way that allows them to feed on scorpions seemingly unaffected by the stings.

DO BATS HAVE ENEMIES?

Whether bats have enemies is easier to document, and the answer is a resounding yes. Not only are bats efficient predators but they are also prey for other predators. Even though they are secretive and nocturnal, and tend to roost in obscure places that give them some protection from their enemies, bats still fall victim to a wide variety of predators. As noted earlier, some are eaten by other, larger, carnivorous species of bats. Other mammalian predators of bats include opossums, raccoons, foxes, skunks, weasels, bobcats, domestic cats and dogs, genets and civets, mongooses, and even some primates, all of which catch the bats while they are roosting. Most take bats from noncave roosts, but occasionally predators also take bats from caves if they roost too far down on the cave walls.

Birds are also bat predators, with birds of prey such as owls, hawks, and falcons of various species taking their toll. Bat falcons are named because of their apparent ability to take bats on the wing. Roadrunners, crows, blue jays, and even blackbirds have been reported attacking bats.

Snakes of many species are known to feed on bats (Figure 1.64), and there are even records of bats being eaten by bullfrogs and trout. Young bats may be eaten by roaches and ants if they fall to the floor of a roost. Spiders may also occasionally take bats. In the rain forests of Panama, I found a large, nocturnal, orb-weaving spider to be an effective predator on the small black myotis, *Myotis nigricans,* I was studying. Even if the spider did not kill and eat the bat but simply cut it out of the web, the bat would probably perish from the sticky web enfolding its wings.

Bats that roost alone or in small groups in the foliage are particularly vulnerable to passing predators such as snakes. Those species that form huge aggregations in caves or mine tunnels face threats from predators that gather at the entrances to their roosts, where they are sometimes forced to exit together through a relatively narrow space. Snakes and small mammalian carnivores may take bats right out of the air in such situations.

Figure 1.64. A gray rat snake, *Elaphe obsoleta*, eating a southeastern myotis, *Myotis austroriparius*, at the entrance to a Florida bat cave. Where caves contain large bat colonies, these snakes gather to feed on emerging bats. They hang from ledges in the cave entrances or from the branches of surrounding trees, waiting for the bats to pass by. Snakes cannot catch a bat unless they are first touched by it. They hang with their noses in the flight path and at the slightest touch of a wing tip quickly strike. One snake may catch up to four or five bats in an evening.

Despite all these natural enemies, bat populations seem to be in no particular danger except where they have run afoul of the most efficient and deadly predator of all, our own species. *Homo sapiens* definitely pose the greatest threat to those bat populations that are in danger of extinction. People kill bats directly because of irrational fear and lack of understanding of their importance to our ecosystems. They harvest them for food in some areas. The biggest threat of all is simply the inevitable encroachment of human populations and their developments, which continually shrink natural habitats.

Subsistence harvesting of large flying foxes for food by native peoples probably had little effect on populations in the past. However, with modern hunting techniques, any market for the animals will quickly lead to serious threats to local populations. Fruit bat is an important ceremonial dish for the indigenous Chamorro people on the island of Guam in the Central Pacific Ocean (Figure 1.65). Their use of these animals has driven the local population on the island to record low numbers, and only strictly enforced protection has saved the bats from extinction. However, the demand has spread to islands as far away as Samoa in recent years, and bat populations on many other islands have been affected by this trade.

Figure 1.65. An endangered Marianas flying fox, *Pteropus mariannus*, feeding on the sweet, ripe fruit of a cycad tree (*Cyas circinalis*) on the island of Guam. Bats appear to be the only native animals that can effectively disperse the seed of a cycad.

If we are going to maintain the 925 species of bats we now share this planet with, we need to learn to live with them in a much less threatening fashion. Bats are so important to so many of our ecosystems that we simply cannot afford to continue to drive them to extinction. The bad press they have suffered in years past is changing rapidly, and perhaps the next generation will grow up with a better appreciation for bats and their worth than we have shown.

DO BATS GET SICK?

Unfortunately, bats do get sick. They are subject to many diseases, as are all mammals, including, or even especially, human beings, and on occasion bats can pass those diseases on to humans. This has been one of the major detriments to bat populations; they are frequently maligned as a health hazard and suffer accordingly.

The overall effect of disease on bat populations is quite difficult to assess, because sick bats are rarely encountered. Bats are frequently associated with histoplasmosis, a fungal disease that causes grave problems in some humans, but there is little evidence that the bats are affected by it. Similarly, many humans carry antibodies suggesting prior exposure to the disease but remain asymptomatic. The fungus grows actively in accumulations of bat guano, particularly in warmer areas. Care should be taken when handling large quantities of guano, as the disease can be quite debilitating if a human inhales large amounts of dust carrying the fungus.

By far the biggest public health concern with bats is rabies, a viral infection of the central nervous system. Like all mammals, bats are capable of contracting rabies. They also suffer the symptoms of the disease and eventually die from it. Bats are not unaffected carriers of rabies. The disease can manifest itself in two quite distinct ways, paralytic rabies and furious rabies. Animals with paralytic rabies become immobilized and may seem less threatening even though they are equally dangerous. Furious rabies is the condition most people are familiar with. Animals with furious rabies may wander dazedly, biting fiercely at whatever they encounter. Bats rarely exhibit this form of the disease and typically will bite only if handled.

There are many strains of rabies in mammals, and the virus seems to move in waves, with various kinds of animals affected in different areas. Any bat that is acting strangely should be avoided. If a bat is on the ground, or otherwise allows you to approach it, leave it alone. The chances of contracting rabies are minuscule in most situations, but handling a sick animal of any kind increases the odds considerably.

The incidence of rabies in bat populations is difficult to measure accurately. Up to 5 to 10 percent of the bats turned in for testing in some localities may be positive for rabies. But this figure is considerably inflated because only sick bats are likely to be captured and tested. The actual incidence is more likely less than 0.5 percent in most areas.

Some bat species are undoubtedly more susceptible to catching and passing on rabies than others. Vampire bats, feeding directly on the blood of other mammals, are far more likely to contract the disease than are fruit-eating bats, for instance. The incidence of rabies in vampire bat populations is probably higher than in almost any other species of bat. There is now a human vaccine for rabies that is both safe and painless, and anyone handling wild animals on a regular basis should obtain preexposure immunization. (See also *Are Bats a Threat to Humans?* and *Do Bats Carry Diseases?* in the section "Bats and Humans.")

A. Chapin's free-tailed bat, *Chaerephon chapini*, Africa.

B. Mexican funnel-eared bat, *Natalus stramineus*, Latin America.

C. Buettikofer's epauletted bat, *Epomops buettikoferi*, Africa.

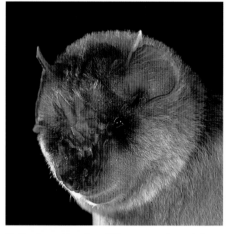

D. Persian trident bat, *Triaenops persicus*, Asia.

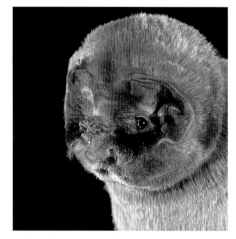

E. Peters's leaf-chinned bat, *Mormoops megalophylla*, Latin America.

F. California leaf-nosed bat, *Macrotus californicus*, North America.

G. Greater bulldog bat, *Noctilio leporinus*,
   Latin America.

H. Australian ghost bat, *Macroderma gigas*, Australia.

I. Commerson's roundleaf bat, *Hipposideros
   commersoni*, Africa.

J. Greater long-fingered bat, *Miniopterus inflatus*,
   Africa.

K. Greater sac-winged bat, *Saccopteryx bilineata*,
   Latin America.

L. Red flying fox, *Pteropus scapulatus*, Australia.

M. Gambian epauletted fruit bat, *Epomophorus gambianus*, Africa.

N. Straw-colored fruit bat, *Eidolon helvum*, Africa.

O. Lander's horseshoe bat, *Rhinolophus landeri*, Africa.

P. Tent-making bat, *Uroderma bilobatum*, Latin America.

Q. Spotted bat, *Euderma maculatum*, North America.

R. Hairy slit-faced bat, *Nycteris hispida*, Africa.

S. Striped hairy-nosed bat, *Mimon crenulatum*, Latin America.

T. Hoary bat, *Lasiurus cinereus*, North America.

U. Tomes's sword-nosed bat, *Lonchorhina aurita*, Latin America.

V. Black flying fox, *Pteropus alecto*, Australia.

W. Big free-tailed bat, *Nyctinomops macrotis*, North America.

X. Long-tongued dawn fruit bat, *Eonycteris spelaea*, Asia.

.2.

BAT EVOLUTION
AND DIVERSITY

WHEN DID BATS EVOLVE?

Bats first appear in the fossil record in the early Eocene epoch, around 60 million years ago. Although in general bat fossils tend to be quite fragmentary because of the fragility of their bones, the oldest fossils known are among the most complete skeletons found to date. These Eocene fossils from Germany (*Archaeonycteris*, *Hassianycteris*, and *Palaeochiropteryx*) and North America (*Icaronycteris*) are beautifully preserved and clearly represent fully formed bats. They are so obviously like modern-day bats that they can readily be assigned to the suborder Microchiroptera. Unfortunately, this tells us very little about the ancestors of bats and what the transition from their terrestrial precursors must have been like.

Because these fully formed bats were around in early Eocene times, paleontologists speculate that bats as a group must have evolved even earlier, perhaps in Paleocene or even late Cretaceous times (70 to 100 million years ago). That bats looked much as they do today at such an early time is astonishing when other kinds of mammals are considered. The first mammals appear in the fossil record only about 200 million years ago, around the boundary between Triassic and Jurassic times. Many ungulates, such as horses and antelopes, underwent major evolutionary changes much later in geologic time; the earliest horses were relatively tiny creatures, quite different from the massive animals we know today. Even our own ancestors were very different in Eocene times, and although we would easily recognize them as primates, they would bear little resemblance to modern-day *Homo sapiens*.

Our knowledge of the evolutionary history of bats is spotty at best. The first fossils were discovered only in the early part of this century. Since then the number of known fossils has increased steadily to about 30 genera in 11 families representing some 40 species. Another 100 or so species of living bats are also known as fossils from the Pleistocene, or the most recent ice age. This gives us reasonably broad coverage, but the fact that most fossils are known from fragmentary remains, and the geographic coverage is essentially worldwide, means that the depth of our knowledge is meager indeed.

Although the first microchiropterans are on the order of 60 million years old, the first megachiropteran fossils are only from the Oligocene, about 35 million years ago. *Archaeopteropus transiens,* from Italy, is a perfectly good representative of the family Pteropodidae, complete with a claw on the second finger, a distinguishing feature of Megachiroptera. Whether or not these earliest megachiropterans also practiced the fruit-eating habits of their modern descendants is unknown.

Because we know so little about the early evolution of bats, ideas about how they made the transition from terrestrial mammals are purely speculative. Although there is no consensus, a possible scenario can be put forth. The common ancestor of the two suborders of Chiroptera was probably a small, arboreal generalist with much less specialized food habits than modern bats of either group. True powered flight was probably preceded by a series of gliding forms, which moved from tree to tree the way modern-day flying squirrels do. Perhaps even better models are the animals we know as flying lemurs, or dermopterans. These large, gliding, arboreal animals are found in the forests of Southeast Asia and bear some physical resemblance to the flying foxes (Megachiroptera) of that region. It is easy to envision an evolutionary progression of increasingly adept gliders, with the gliding membranes becoming expanded between the finger bones. Mammalian forelimbs require little modification to be able to provide power to wings, and the appropriate muscle masses could have evolved in synchrony with the other changes.

If such a scenario is correct, the Megachiroptera and Microchiroptera must have split apart early in that history. Microchiroptera have dental patterns that readily group them together and that suggest insectivory as the primitive stage. Megachiroptera, by contrast, have rather bizarre teeth that are not readily derived from the primitive insectivore pattern.

One theory that has waxed and waned over the last few decades would solve the problem by suggesting that Megachiroptera and Microchiroptera are not necessarily each other's closest relatives. Some recent evidence has found neural pathways in the brain suggesting that Megachiroptera are more closely related to Primates than to Microchiroptera. The difficulty with this theory, aside from the fact that little additional evidence has been found to support it, is that it would require the evolution of wings independently in the two groups. Given the overwhelming similarities of the groups, independent evolution of flight seems unlikely at best. Also, recent studies of molecular biology have consistently supported the more conservative view of common ancestry between the two suborders of Chiroptera.

Figure 2.1. Bat biologists examining cave deposits in Africa. Such areas occasionally yield fossilized bits of bat bones and teeth that can provide valuable clues to bat evolution.

WHERE ARE FOSSIL BATS FOUND?

Bat fossils are amazingly well distributed around the globe. They are now known from all major continents: North and South America, Europe, Asia, and Africa. Bat fossils are frequently found in cave deposits, not surprising considering modern species' propensity for roosting in caves. Such deposits often contain considerable amounts of material that has been reduced to an almost rubblelike state. This material can be sifted for teeth and in some cases bits of jaw, both relatively durable parts of the skeleton, which can be identified by painstaking paleontological work (Figure 2.1).

The complete skeletons that have been found are in highly fossiliferous shale beds. Such preservation is rare, and although the fossils are from widely scattered geographic localities, they are few in number. A more promising source of additional material is careful reanalysis of some major cave faunas, where additional postcranial elements are occasionally found. Such deposits, if they can be positively tied to currently recognized groups, may lead to considerable information on past climates and ecosystems.

HOW MANY SPECIES OF BATS ARE THERE?

The most recent compilation of currently recognized species of mammals in the world lists 925 species of bats. Since that list was published, in 1993, additional species, primarily from South American rain forests, have already been described.

About 30 new species of bats have been described since 1985. Other species remain to be discovered, and the classification of bats will continue to be improved by bat taxonomists, or systematists. Bats are not a particularly easy group to study because of their secretive, nocturnal habits and their flight capabilities.

One entire family of bats, the Craseonycteridae, was described as recently as 1974. These bats are potential claimants to the title of world's smallest mammal, at about 2 grams (see Figure 1.1). Their size along with their restricted distribution in a limestone-rich area of Thailand are probably what allowed them to live undetected for so long. Most discoveries are not so dramatically different from their relatives. Newly described species frequently differ in relatively minor ways from species already known to exist in other areas. However, recent techniques of molecular biology are helping to refine systematists' ability to distinguish between closely related species of bats and other organisms.

Although new species are constantly being discovered and described, the number of recognized forms can also fluctuate downward. Bat systematists are constantly refining classifications as new knowledge is gained, and sometimes two kinds that had previously been recognized as separate species are lumped together, with one name becoming a synonym for the other. According to a carefully documented code of scientific nomenclature, the oldest name takes precedence.

Many species of bats are still known to science by only a single preserved specimen. Until additional specimens are collected from several localities, it is extremely difficult to judge their true relationships. That means that bat biologists are still actively engaged in the collection and preparation of scientific specimens from around the world. Such material is indispensable in determining evolutionary history.

WHAT CHARACTERIZES THE MAJOR GROUPS OF BATS?

The following sections characterize the major groups of bats by selecting representative examples of each group. For a complete listing of the genera and species in each of the families or subfamilies discussed here, see the appendix. The order Chiroptera is divided into two suborders, Megachiroptera and Microchiroptera.

The suborder Megachiroptera, with the single family Pteropodidae, contains the animals called flying foxes, Old World fruit bats, or megabats. They have undergone a long and distinct evolutionary history apart from their sister group, the Microchiroptera, or microbats, but all are easily recognizable as bats. Megachiroptera are so different from Microchiroptera in some characteristics that some biologists have suggested that they are more closely related to Primates than to Microchiroptera. The single most readily discernible morphological difference is the presence of a claw on the second finger in megabats, a trait lacking in microbats. The

Figure 2.2. A Pacific flying fox, *Pteropus tonganus*, roosting in the top of a tree with its wings wrapped around itself.

rapidly building consensus, based on a variety of recent molecular studies, seems to be that megabats did share a common ancestor with microbats more recently than either did with any other major group of mammals.

The most significant behavioral difference between the two groups is megabats' lack of reliance on echolocation. They are much more visually oriented than microbats, although a few species have a relatively inefficient, low-frequency system of echolocation based on tongue-clicking sounds. Megabats are much less diverse in food habits than microbats; all megabats feed on either fruit or flowers, including both nectar and pollen.

Most of the large species, such as those in the diverse genus *Pteropus*, form large colonies, known as camps in some areas, in the tops of trees. They roost in the open in the canopy, often occupying many adjacent trees. In Fiji we found a colony of several thousand Pacific flying foxes, *Pteropus tonganus*, spread over nearly 15 trees, many of which were showing serious wear and tear from the constant comings and goings of these bats (Figure 2.2).

Members of the genus *Rousettus*, with their rudimentary echolocation systems, are able to occupy cave habitats in much the way many microbats do. This habit, in turn, occasionally allows for local densities of rousette fruit bats to be quite high. In the summer of 1996, I captured and released dozens of *Rousettus leschenaulti* (see Figure 1.8), which were easily the most common bat species in the Chatthin Wildlife Sanctuary in north-central Myanmar.

Flying Foxes

Flying fox is an English common name that refers in general to the entire suborder Megachiroptera and its only family, the Pteropodidae. However, the group is probably better referred to as Old World fruit bats. Even this is a bit of a misnomer, because there are several species that feed on nectar as well as fruit. Flying foxes, however, is sometimes used more specifically for the 58 species in the genus *Pteropus*, which includes the largest known species of bats (Figure 2.3).

These are spectacular animals, roosting in the tops of canopy trees, active and noisy during the daytime and very impressive against the sky when in flight. They form huge colonies in some areas and are important dispersers of seeds in many parts of Asia and on Pacific islands. Some species are nomadic, moving through areas in search of fruiting or flowering trees. Others make long-distance commutes between day roosts and nighttime foraging areas (Figure 2.4).

The flying foxes are so named because of their resemblance to canids, with their long muzzles and large eyes. They are visually orienting and probably find their food by olfaction. They feed mainly by crushing fruit against their bony palates and using their well-developed teeth in chewing, then swallowing the juice and some

Figure 2.3. (left) Lyle's flying foxes, *Pteropus lylei,* one of the largest species of bats, roosting in a treetop in Southeast Asia.

Figure 2.4. (right) A solitary Pacific flying fox, *Pteropus tonganus,* flying majestically over a Samoan rain forest.

pulp, and spitting out seeds and most of the pulp. Some seeds are invariably ingested, and material goes through the digestive tract quite rapidly, dispersing the seeds throughout the foraging area. Some species hover and bite into large fruits; others land and manipulate the fruit with feet and wrists as well as teeth; still others pluck fruit and fly to a separate feeding roost to process it. These bats are also important dispersers for fruits with a single large seed in the center, such as cycads, which are carried away and dropped (see Figure 1.65). Some flying foxes have lived in captivity for 25 years.

Blossom Bats

One subfamily of the Pteropodidae, the Macroglossinae, contains six genera whose species are in general much smaller than their relatives in the other subfamily, the Pteropodinae. Macroglossines are pollen and nectar feeders by and large, although almost all take fruit on occasion as well. Blossom bats belong to the genus *Syconycteris*, which contains three species in Australia and New Guinea, and on nearby islands (Figure 2.5). They are among the smallest megabats, smaller than many microbats and about the same size as most common bats in North America.

Figure 2.5. A southern blossom bat, *Syconycteris australis*, pollinating a swamp banksia plant.

Figure 2.6. Banana bats, *Pipistrellus nanus*, roosting inside the unfurled leaf of a banana plant.

Although the megabats are a rather homogeneous lot when it comes to food habits, the microbats show an amazing diversity. All the food habits known in bats are found in the suborder Microchiroptera, including fruit, nectar, pollen, insects, vertebrates, fish, and blood. To pursue these differences in prey items, microbats employ a concomitant diversity of flight styles, ranging from slow, fluttering, and maneuverable to fast and direct. These differences are afforded by distinct variations in wing shape and size, and the related measures of wing loading and aspect ratio.

Microbats' eyes are small and in some cases almost hidden. All microbats rely heavily on echolocation for navigation, orientation, and prey detection. Many of the behavioral adaptations of echolocation are accompanied by morphological characteristics ranging from ear size to facial adornments such as flaps, folds, and nose leaves. Teeth are complex in most species, consisting of a relatively ancient and primitive pattern of W-shaped cusps that allow for cutting, slicing, and grinding of food items.

Microbats frequently roost in dark caves, mine tunnels, or other retreats, such as tree cavities and attics of buildings. Some also roost in the foliage of trees or bushes but never in the open as do megabats. Some microbats have highly specialized adaptations for modifying vegetation into "tents," and some have adopted very limited roosting sites, such as the insides of rolled leaves or the nodes of bamboo stems (Figure 2.6).

Mouse-Tailed Bats

Mouse-tailed bats belong to the family Rhinopomatidae, which contains only a single genus, *Rhinopoma*, and three species (Figure 2.7). They occur in Asia, across the Middle East, and in Africa. All three are relatively small bats, with long tails that are nearly as long as the heads and bodies combined. The tail extends well beyond the interfemoral membrane, which is quite short. Mouse-tailed bats have handsome faces, with large ears connected by a low band of skin across the forehead and a tiny nose leaf on the tip of the snout.

Figure 2.7. A Hardwicke's mouse-tailed bat, *Rhinopoma hardwickei,* roosting on a cave wall.

They occur in arid regions, where they roost in caves, cliffs, wells, houses, and even the pyramids in Egypt. Some are known to store large amounts of fat and to enter torpid periods in the winter that are not quite hibernation. Nevertheless, they can sometimes go for several weeks in captivity without food or water, living on their fat stores. They may form colonies numbering in the thousands in suitable roosts, but others roost in groups of 4 to 10.

Hog-Nosed Bats

The hog-nosed bat, the world's smallest known mammal, was first described in the mid-1970s. It is still known only from a few caves in Thailand and represents the sole member of the family Craseonycteridae. *Craseonycteris thonglongyai,* named for an excellent student of bats from Thailand, Kitti Thonglongya, is about the size of a large bumblebee, with a head and body length of about 30 millimeters and a weight of about 2 grams (see Figure 1.1).

Not only are hog-nosed bats the smallest bats but they are also among the least known. Insectivorous, they have been observed feeding around bamboo clumps and teak trees. Most of the forest in the immediate area of the caves where these bats roost has been cleared, which probably does not bode well for their future.

White Bats

The genus *Diclidurus,* a New World member of the family Emballonuridae, contains four species that are among the very few all-white bats. Appropriately, they are sometimes called ghost bats. Actually, their hairs are grayish at the base and white only on the outer half or so, but these bats' overall appearance is distinctly white.

One species, *Diclidurus albus,* the northern white bat, is known from Mexico and Central America and extends southward to Brazil. These bats roost in the foliage of palm trees in Mexico, and they feed by catching insects on the wing while flying high above the forest canopy. The only other white bat known is the white fruit bat, *Ectophylla alba,* a member of the family Phyllostomidae, restricted to a narrow band of rain forest habitat along the Caribbean slopes of Central America.

The family Emballonuridae is characterized by a tail that is shorter than the uropatagium, and the tip of the tail protrudes through the upper surface of the tail membrane (see photo *K, Saccopteryx bilineata,* in the gallery). As a result, one common name for the whole family is sheath-tailed bats. Other members of the family have glandular pouches in their wings, resulting in another popular common name, sac-winged bats.

Slit-Faced Bats

Slit-faced bats, also known as hollow-faced bats, have a furrow along the top of the muzzle from the nostrils to a pit in the middle of the forehead. Nose leaves and fur, which conceal the hollow externally, lend an interesting appearance, but the slit is readily apparent in the skulls of these bats.

The family Nycteridae contains only the genus *Nycteris*, which has 12 species with distribution limited to parts of Africa and Asia (see photo *R*, *Nycteris hispida*, in the gallery). They occupy both forest and savanna habitats, and feed on large insects that they capture by gleaning from the surface of the vegetation or the ground. The Egyptian slit-faced bat, *Nycteris thebaica*, feeds on scorpions, much as does the American pallid bat, *Antrozous pallidus*.

Most species roost in small groups in caves, hollow trees, foliage, rock outcrops, or even occasionally in the burrows of larger mammals, such as porcupines or aardvarks. Other species form aggregations of several thousand individuals in caves.

Old World False Vampire Bats

The family Megadermatidae contains five species in four genera, all restricted to the Old World. *Macroderma gigas*, the Australian ghost bat, is the largest member of the Microchiroptera (see photo *H* in the gallery). Their forearm length is over 100 millimeters, they weigh over 100 grams, and their wingspan is over half a meter.

These bats are striking animals, with long, erect ears and a nose leaf. They are a pale, ashy gray with whitish head, wings, and underparts. This coloration has led to the common name of ghost bat for this species, as well as false vampire bat, demonstrating the problems caused by referring to animals by their common names. Both common names are misleading, because these bats do not feed on the blood of other species. Some are carnivorous and prey on a variety of small vertebrates, but they consume the entire animal, just as do many other carnivores.

Macroderma gigas are known to feed on small mammals, which they capture by dropping on them from above, wrapping their wings around them, and biting them in the neck and head. They are fond of house mice but also take other rodents, small marsupials, smaller bats, birds, reptiles, and large insects.

These bats roost alone or in small groups in caves and mine tunnels. Apparently they used to be much more common in Southern Australia, where they no longer occur. They are threatened by human activities that damage their roosting sites and may be limited by changing climatic conditions as well. They are considered vulnerable by the International Union for the Conservation of Nature (IUCN).

Horseshoe Bats

The genus *Rhinolophus* contains 64 species widely distributed throughout the Old World. Their name comes from a complex nose leaf made up of three parts, the lowest of which is horseshoe shaped. The upper part is an erect spear, similar to that of the New World leaf-nosed bats in the family Phyllostomidae. Another similarity is the emission of ultrasonic sounds through the nose instead of through the mouth, as in most other bats. This leads to the obvious speculation that the nose leaf is somehow involved in directing the ultrasonic emissions.

These bats' echolocation system is distinctive in having a long narrowband component thought to be particularly useful in detecting fluttering insects. Their ears are large and lack tragi. These bats frequently forage fairly close to the ground, and some species may take insects or spiders directly from the ground.

Their roosting posture is also distinctive, in that the wings are frequently wrapped around the body (see Figure 1.37). When hanging upside down in this fashion, these bats resemble large cocoons. They roost in caves, buildings, hollow trees, and sometimes directly in foliage. Some species are solitary, but others may form fairly large colonies in the proper roosts. Temperate zone species hibernate during the winter. They are known to live for at least 24 years, based on a banding recovery in Germany.

All species studied to date have only a single young, produced after a gestation period of about seven weeks of active development. In the hibernating forms mating may occur in the fall, with fertilization arrested until the spring. Lactation lasts for four to six weeks, and the young reach sexual maturity in two years.

Roundleaf Bats

Along with the subfamily Rhinolophinae, the Hipposiderinae compose the family Rhinolophidae. Although there are nine genera in the subfamily, the genus *Hipposideros*, with 53 species, is the most widespread and abundant. They are found throughout the Old World tropics in Africa, Asia, and Australia. As their name suggests, they also have an elaborate, rounded nose leaf and emit their echolocation calls through the nose.

Both subfamilies are also known as fossils from the Eocene period in Europe. Hipposiderines differ from rhinolophines in details of nose leaf morphology, foot morphology, dentition, and form of the shoulder and pelvic girdles. Their feet are unusual in having only two bones in each toe, as opposed to three in each except the first toe in rhinolophines. Although most of these bats are small to medium size for microchiropterans, Commerson's roundleaf bat, *Hipposideros commersoni*, is one of the largest microbats, with a forearm length of over 100 millimeters (see photo *I* in the gallery). Most hipposiderids are brown or reddish in color, and some individuals are even a rich burnt orange. Their ears are large and funnel shaped and lack tragi. The short tail, when present, is enclosed in the interfemoral membrane.

Hipposideros roost alone or in colonies of up to several thousand individuals, depending on the species. They use caves or tunnels for the most part, although some are found in hollow trees, buildings, or even in the underground burrows of large mammals. In Africa, Sundevall's roundleaf bat, *Hipposideros caffer*, may form colonies of up to half a million.

These bats are strictly insectivorous, and they catch much of their prey on the wing. They feed on beetles, termites, cockroaches, and cicadas, among other things. *Hipposideros commersoni* have been reported to feed on the larvae inside wild figs.

Reproduction appears to be seasonal, and most roundleaf bat species studied to date have only a single young per year. As with most insectivorous bats, their reproductive season seems to be timed so that the young are weaned during the period of greatest insect abundance. This means that the cycles vary geographically, depending on rainfall regimes.

Bulldog Bats

The family Noctilionidae contains only the genus *Noctilio,* with two species. The greater bulldog bat, *Noctilio leporinus,* is also known as the fishing bat, because of its fish-eating habits (Figure 2.8). These bats' head and body length is over 100 millimeters, and they weigh around 75 grams. Their hind feet and claws are tremen-

Figure 2.8. A greater bulldog bat, *Noctilio leporinus,* resting on a tree trunk, with fish in its cheek pouches.

dously enlarged. Coloration ranges from bright orange rufous to a dull, pale gray. There is a well-marked middorsal stripe in both sexes.

These bats' long hind legs support a well-developed uropatagium, which must be lifted high above the water surface to avoid friction. The lifting is done by greatly enlarged calcars, the small bones extending from the ankles and providing support to the trailing edge of the membrane. The tail is well developed and extends about halfway out the membrane, where the tip may protrude through the dorsal surface in a sheath.

Their wings are quite long and broad, and provide sufficient lift at slow speeds to allow the bats to carry a small fish up and away from the water surface. The long rostrum and nose lack any type of appendage. Their cheeks are expandable, permitting the bats to fly with a considerable amount of food in their mouths. Their ears are long and pointed, with tragi.

Noctilio leporinus commonly roost in rock cliffs and fissures, as well as caves and tunnels, and they have been taken in hollow trees and occasionally in buildings. Both species of bulldog bats produce a strong and distinctive musky odor, at least partially from bacteria-filled glands around the anus.

These bats catch small fish in streams, lakes, and even over the sea. They use echolocation to detect either surface ripples or the fish themselves. They skim the surface of the water, trailing their long claws in the water, and gaff the fish. They usually transfer the fish to their mouths, where their distendable lips help surround the prey, and the bats either eat on the wing or carry their prey items to a feeding roost, where they eat while hanging. If they should get dumped accidentally into the water, these bats swim quite well and work their way promptly to shore, where they clamber up the nearest object to gain elevation for getting airborne once again.

Southern bulldog bats, *Noctilio albiventris*, feed on insects, although they have been induced to eat fish in captivity. All the individuals I have captured in Central America have had only insect remains in their stomachs or feces. They feed in large swarms very near the water, suggesting that they are catching insects just above the surface.

The bulldog bats' reproductive cycle is thought to be monoestrous, which means they breed only once per year and produce only a single young per pregnancy. The records are quite scattered, however, and year-round data are lacking for both species.

Noctilio is one of only three bat genera known to feed on fish. The other two are *Myotis* and *Megaderma*. Fish eating is a specialized form of carnivory, but it is not so different from gleaning foliage for large insects. Not only is this a curious specialization for bats but these are spectacular bats. Zigzagging just centimeters above the water surface, they dip suddenly, and their claws make a hissing noise as they hit the water six or seven times in quick succession. A successful capture results in a small splash and the bat's instantaneous upward flight.

Naked-Backed Bats

The family Mormoopidae has only two genera and is limited to the New World (see photo E, *Mormoops megalophylla*, in the gallery). One genus, *Pteronotus*, has several species, two of which are naked-backed bats, so called because their wing membranes meet and fuse in the middle of the back, giving the bats a smooth, naked-backed appearance. Underneath these membranes are normally furry bodies. None of these bats is very large, but Parnell's mustached bat, *Pteronotus parnellii*, is the largest, with a forearm length of about 60 millimeters.

These bats have no nose leaf, but their lips are wrinkled and modified into a sort of funnel-shaped appearance. Their eyes are small and quite inconspicuous. Their ears are short and pointed, and the tragi are not only present but quite distinctive in comparison to those of other bats. Their interfemoral membrane is well developed, and the tail is always present and extends through the dorsal surface of the uropatagium.

These bats are agile flyers and are strictly aerial insectivores. They occur in a variety of habitats, ranging from dry savannas to tropical rain forests. Most roost in caves or tunnels, although they have been found in houses on occasion. They produce a single young per year in seasonally determined patterns.

Spectral Vampire Bats

Vampyrum spectrum—the name alone conjures up bloodthirsty images that are quite removed from the reality of these remarkable bats. They are not vampires at all but rather carnivores. Members of the large New World family Phyllostomidae, spectral vampires are the largest bats in the Western Hemisphere. Their head and body length is about 130 millimeters, their forearms are 100 millimeters, and wingspan reaches nearly a meter in the largest specimens.

Vampyrum are top carnivores among bats, feeding on birds, other bats, and rodents. They forage for prey items that they can take either to a feeding roost or back to their main roost to eat. This is one of the few species of bats known to provision their young by bringing prey back to the roost. They are easily kept in captivity in zoos and will take a variety of food items, including frogs, lizards, and large insects in addition to small birds and mammals.

I have captured male and female pairs on several occasions, suggesting that they may forage together, or at least fly together at some point during the night. They may be attracted by the calls of other bats in nets, or perhaps they are drawn to bats being held in cloth bags near the nets. When held these bats make an audible sound not unlike purring, and they constantly cast their large ears forward and from side to side, as though surveying incoming sounds or echoes.

Although they occur from Mexico to Brazil, these bats tend to live in low population densities and seem to prefer primary forest to other types of habitats. Each female gives birth to only a single young per year.

Figure 2.9. A fringe-lipped bat, *Trachops cirrhosus*, and a Linnaeus's short-tailed bat, *Carollia perspicillata*, entering a hollow tree in Latin America. Both are members of the subfamily Phyllostominae.

Vampyrum are another example of tropical American bats inappropriately named by a taxonomist working in a European museum. In fact, they were named by Linnaeus himself, among seven bats he recognized in his tenth edition of *Systema Naturae*, published in 1758. At that time vampire bats were known to occur in the Americas, and new specimens from that part of the world were frequently given names that implicated them as vampires.

Vampyrum are members of the New World leaf-nosed bats, the family Phyllostomidae and the subfamily Phyllostominae. This group of 11 genera and 33 species represents a relatively old lineage of bats that share general characteristics of large ears and nose leaves, broad wings that produce slow, maneuverable flight, and a propensity for life in old, well-developed forests (Figure 2.9). They tend to prey on larger insects that they frequently take from the surface of vegetation.

Nectar-Feeding Bats
The phyllostomid subfamily Glossophaginae includes 10 genera and 22 species. Most are highly specialized for feeding on nectar. They tend to have elongated snouts, long, extrusible tongues, and reduced dentition. Their ears are short, their

wings are broad, and most are capable of highly maneuverable, even hovering, flight (Figure 2.10).

Leptonycteris curasoae, the southern long-nosed bat, is a representative member of this group with an interesting lifestyle. These bats are known from the southwestern United States, where they are summertime residents only, to northern Central America, and also in northern South America and nearby islands. The northern populations are migratory and appear to follow the flowering cycles of large *Agave*, or century plants (see Figure 1.63).

The bats arrive in southern Arizona in late spring or early summer and leave in late summer or early fall. While in residence, in addition to feeding on flowering agaves, saguaros, and organ-pipe cacti, they tend to visit hummingbird feeders. Residents in the Chiricahua and Huachuca Mountains of southern Arizona began to notice some years ago that the sugar water in their hummingbird feeders was disappearing overnight, when hummingbirds were inactive. A little nighttime detective work with a flash camera produced incontrovertible evidence that nectar-feeding bats in the genera *Leptonycteris* and *Choeronycteris* (Mexican long-tongued bats) were the culprits.

Leptonycteris are a valuable resource in Mexico, where they pollinate large agaves and cacti (Figure 2.11). These include the huge saguaro-type cardon cacti and the commercially important magueys that are used to produce tequila. The next time

Figure 2.10. A nectar-feeding Underwood's long-tongued bat, *Hylonycteris underwoodi,* pollinating a tropical flower.

Figure 2.11. A southern long-nosed bat, *Leptonycteris curasoae,* **(A) approaching a saguaro cactus (***Carnegiea gigantea***) flower, attracted by the bright white petals and the sweet-smelling nectar at its base; (B) lapping nectar in the bottom of the flower; and then (C) withdrawing its pollen-covered head to forage elsewhere.** Like a key and a lock, this bat's head and this plant's flower are closely matched. The pollen that sticks to the foraging bat is transferred to other flowers as the bat makes its rounds. A single bat may visit as many as 100 cactus flowers each night. Each pollen grain initiates the development of fruit and is destined to become a viable seed. In areas where bats are scarce many cacti go unpollinated and these plants may be yielding fewer seeds than necessary to ensure their replacement. Such a shortage could have far-reaching effects on the entire Sonoran Desert ecosystem. Many other animals, including day creatures, rely on the cacti for food and shelter. Small mammals and birds seek fruit or shelter among the spines of these majestic plants. Many animals, including other species of bats, live in holes carved out by woodpeckers. Not only do bats, birds, and insects gain precious moisture as well as energy from the nectar of the cactus flowers, but also mammals such as antelope ground squirrels, foxes, and ringtails feed on the fruit.

A.

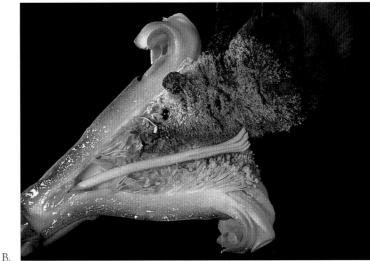

B.

C.

you enjoy a margarita or tequila sunrise, you might want to raise a toast to the hardworking *Leptonycteris*, who are providing the nightly pollination services that make our lives that much more enjoyable!

Flower Bats

Another group of nectar-feeding bats are the flower bats of the Caribbean. These three genera and five species are limited to Caribbean islands, where they play a role in pollination for a variety of species of flowers similar to that of the glossophagines and lonchophyllines on the mainland.

Brachyphylla cavernarum, the Antillean fruit-eating bat, is a common species on Puerto Rico and the Virgin Islands, occurring south through the Lesser Antilles to St. Vincent and Barbados. The scientific name is a mouthful, but it is appropriately descriptive of both the morphology and lifestyle of this species. *Brachyphylla* means "short leaf," referring to the vestigial nose leaf, and *cavernarum* means "cave dwelling," which is where the species lives for the most part, although these bats do occasionally roost in buildings as well.

In addition to nectar and pollen, these bats appear to be opportunistic feeders on a variety of items, including fruit and even insects on occasion. They form colonies of several thousand individuals in appropriate caves.

Long-Tongued Bats

Three genera and nine species of long-tongued bats compose the subfamily Lonchophyllinae. This group, closely related to the Glossophaginae discussed in the previous section, is the third major group of pollen- and nectar-feeding bats. Superficially, they are also quite similar to the other nectar feeders, with elongate faces, long tongues, and reduced teeth.

Few details on the natural history of this group are known. The genus *Lonchophylla* is the most diverse and widely distributed, with seven species scattered through northern South America and extending into southern Central America. These bats are all small and restricted to areas that produce ample numbers of flowering plants throughout the year to provide a constant food source.

Short-Tailed Fruit Bats

The subfamily Carolliinae includes two genera and seven species. One of these, *Carollia perspicillata*, is one of the better-known species of tropical bats, thanks to the long-term studies of Ted Fleming of the University of Miami (see Figure 1.56). He spent ten summers studying this species in the dry forest of Costa Rica, and his detailed book on their life history provides a comparative basis for all such studies in the future.

This species is widely distributed from Mexico to Argentina, and these are also among the most common bats in suitable habitat throughout that range. This combination of being widespread and common is unusual, but it does occur with a few other bat species. The Jamaican fruit bat, *Artibeus jamaicensis,* and the long-tongued nectar bat, *Glossophaga soricina,* fit this pattern as well. Those two species, along with *Carollia perspicillata,* are frequently found in second-growth forests, or areas where the original forest has been cut and a regenerated forest has developed in its place.

The habitats favored by the short-tailed fruit bat include far more than second growth, however. Throughout much of their range, they feed extensively on the fruits of *Piper* plants, a genus that includes commercial black pepper. In some areas these are pioneer species, among the first to colonize openings of any sort, including tree-fall gaps, streamsides, forest edges, and even regenerating fields.

These bats have a bimodal reproductive cycle, with each female producing a single young in each of two birth periods per year. In Costa Rica the first of these occurs in the dry season, resulting in a flush of weanling bats early in the rainy season, when many of their favored food plants are bearing fruit. The second birth peak occurs in midrainy season, still a favorable time of year for bat foraging. Newborns weigh about 5 grams, almost 30 percent of the mothers' body weight.

As do other bats, mother short-tailed fruit bats leave their young behind when they are foraging. Unlike some others, however, they do not always leave them in large groups. They may in fact leave them singly, either in the day roost or in some other spot, perhaps a night roost in the foliage. I first became aware of this possibility some years ago while working in Ecuador. A herpetologist colleague from the National Museum of Natural History, Roy McDiarmid, was searching for frogs along a small stream late one night, and, in looking carefully through the vegetation, he came upon a small bat. He plucked it from its roost and placed it in a cloth bag, which he then carried attached to his belt. After having worked his way down the stream for some distance, he was surprised to find flying around him another bat, which eventually landed on the cloth bag! Being an experienced field biologist, he found this activity far more interesting than alarming and quickly added this bat to the bag.

The next morning he passed the bag to me along with an account of how its inhabitants came to be there. When I opened the bag, I found one bat, presumably the first, nursing from the other, presumably the second. We deduced that the mother had parked the youngster on the branch while she shopped along the stream bank for available fruit and then either found it missing upon her return or intercepted Roy and the bag on her way back. Apparently, the youngster was calling (inaudibly but probably ultrasonically), and the cries attracted Mom directly to the bag.

I have had other experiences of bats attracted to a cloth bag full of bats, outdoors and even inside, where an open window will frequently allow bats to zoom into the room to check out a bag with other bats inside. This further confirms the idea that social communication is important to these animals and underscores how little we understand about it.

Young *Carollia perspicillata* develop rapidly, their forearms almost doubling in length in about six weeks. They are flying by three weeks of age and quite accomplished at it by the time they reach adult size.

The social system of these bats includes harems, with a single adult male guarding a cluster of adult females. Roosts also frequently contain bachelor males and subadult animals of both sexes. The males compete for access to these groups, perhaps through guarding and controlling a particularly favorable site within the roost.

In addition to *Piper*, these bats take a wide variety of other fruits, depending on site and season. They may concentrate on fruits higher in protein early in the evening, then switch to energy-rich fruits in later foraging bouts. They are familiar with their foraging territory and usually feed within a kilometer or two of their day roost. Typically, they pluck a fruit and fly with it to a night feeding roost, where they hang upside down and eat it. This procedure probably helps them avoid predators, which might otherwise be attracted to fruiting trees containing a large number of feeding bats.

American Epauletted Bats

American epauletted bats are sometimes placed in their own subfamily, Sturnirinae, which contains only the single genus *Sturnira*, with 12 species ranging from Mexico to Argentina. However, they are probably more properly considered members of the largest phyllostomid subfamily, the Stenodermatinae. In a number of ways this is an enigmatic group in spite of their relative abundance in many areas. The yellow epauletted bat, *Sturnira lilium*, is common, widespread, and abundant (Figure 2.12). Others are rare, have limited distributions, and are never found in high numbers.

These are fruit-eating bats, and although they may be found in older, primary forests, both in the lowlands and well into the highlands in many areas, they are most abundant in certain second-growth habitats where fruit is plentiful. They roost in hollow trees and occasionally in buildings. In any given region they seem to have the bimodal reproductive pattern common to many phyllostomids. Details of their food habits are not well known.

These bats are readily recognizable by the yellowish spots on their shoulders; hence they are called epauletted bats. The spots are actually glandular areas that are much better developed in males than in females and presumably have some

Figure 2.12. A yellow epauletted bat, *Sturnira lilium*, in flight with a *Solanum* berry in its mouth.

function in reproductive or social behavior. These glands are similar to those of a group of African megachiropterans in the genus *Epomophorus* (epauletted fruit bats), which occupy a similar ecological niche on that continent (see Figure 1.39).

Tent-Making Bats

Tent-making bats belong to the subfamily Stenodermatinae, the largest of the phyllostomid subfamilies, in terms of both number of genera, with 17, and number of species, with 62. Although several species in more than one genus make tents, the best known of these is *Uroderma bilobatum* (see Figure 1.24). The species name *bilobatum* refers to this species's characteristic two-lobed upper incisors.

The species occurs from tropical Mexico through Central America and most of South America to southern Brazil. It is widespread and common in lowland forests of all types. Individuals have been taken as high as 1,800 meters in Peru.

Their common name refers to these bats' habit of modifying leaves into tent roosts by biting and chewing the midribs and veins so that the leaves droop in a characteristic fashion. These tents provide shelter for single individuals or small groups in most instances, but up to 60 animals have been reported from a single tent. Tents are made from different species of plants, including palms and epiphytes. This peculiar habit was first noticed and described by the naturalist Thomas Barbour more than fifty years ago.

Vampire Bats

Probably no other bat carries as much name recognition as the vampire, yet most people understand little about the real vampire bat (see Figure 1.11). Actually, there are three species of vampires, although only the common vampire bat, *Desmodus rotundus*, is common and widespread. All three are limited to the New World tropics, and they have nothing to do with the vampire legends of Transylvania.

Desmodus rotundus occurs from northern Mexico to northern Argentina and central Chile. These bats can be extremely abundant in areas of intensive domestic livestock production, but they are quite uncommon in undisturbed wilderness areas. These are not large bats; adults average about 80 millimeters in length and weigh 30 to 35 grams. They are well adapted for terrestrial locomotion and have exceptionally long thumbs for bats.

They have only 20 teeth, and the only really functional ones are the two chisel-shaped upper incisors and the greatly enlarged upper canines. Their wings are broad, and these bats are capable of taking flight from the ground by leaping. They frequently fly low and can maneuver easily through the understory.

Roosts of vampires are found in caves, mines, hollow trees, and occasionally abandoned buildings. Vampire roosts are easily recognizable by the dark stains of digested and excreted blood on the walls and floor. The blood results in a characteristic pungent odor as well.

Vampires forage early in the evening, usually traveling no farther than 5 to 8 kilometers from their roost in search of prey, although foraging distances of up to 15 to 20 kilometers have been reported. In most areas their normal prey is cattle, horses, or burros, but all large mammals, including humans on occasion, are potential prey. Once a prey item is located, the bat may spend some time approaching it, often from the ground. Next a site is selected, and the bat begins by licking the site with its tongue. In addition, hair or feathers (vampires also occasionally feed on birds) may be removed, and a bite is made, either by a slash with the upper teeth or by nipping off a small piece of skin.

The bat's saliva contains anticoagulants that prevent the blood from clotting. The bat may feed for 10 minutes to half an hour from the same wound. It runs its tongue in and out of the wound, and two small grooves on the underside of the tongue convey the blood to the throat. After a prolonged feeding bout, the bat may become visibly distended.

Although they rarely feed on humans, these bats do have the curious habit of returning to a victim after a successful feeding bout. There are numerous accounts of people sleeping in the same room with the same individual bitten on successive nights while his or her companions were ignored. The amount of blood lost is not detrimental to the prey, human or otherwise, but there are dangers of secondary infections and the transmission of diseases, including rabies. Vampires are more likely to transmit rabies than other species of bats because they come into regular contact with so many other mammals.

How did such a feeding habit come into existence? No one knows for sure, but various theories have been suggested. Vampires are closely related to the basically frugivorous phyllostomids, so one theory has them originating from fruit-eating ancestors. Their short muzzles and strong incisors are equally well suited for nicking vertebrate prey or for ripping into large, thick-skinned fruits. The seemingly necessary intermediate step between feeding on a large, thick-skinned fruit and slicing open a cow, however, is more difficult to imagine.

Another possibility is that foliage-gleaning bats selected large ectoparasites, such as ticks, from the backs of large mammals. This would derive them from insectivorous ancestors and would require an equally difficult transition from feeding on blood-engorged parasites such as ticks to tapping their sources.

The third theory suggests that these bats began to feed on the larvae of insects, such as screwworms, that are common in open wounds of large mammals. This habit would also require the well-developed anterior dentition that could logically be derived from phyllostomidlike ancestors. It presents a somewhat easier to imagine intermediate scenario, of moving from feeding on the insect larvae through feeding on the wounds themselves and eventually to inflicting the wounds in order to feed on the blood alone.

All these theories are highly speculative; we have little or no evidence to support any of them. Such evidence is extremely difficult to gather, as is the case with most studies of evolutionary pathways. We may never know exactly how this bizarre behavior developed, but as long as scientists allow their curious minds to roam over the possibilities, there is the chance that someone will eventually disprove all but a single, viable hypothesis.

Regardless of the ultimate origin of their feeding habit, vampire bats present a fantastic example of the specializations provided by the evolutionary process over a long period of time. Even without the vampire legends from the Old World, these animals are truly amazing.

Funnel-Eared Bats

The Mexican funnel-eared bat (*Natalus stramineus*), the most widespread member of this single-genus family restricted to the New World tropics, is a small, delicate bat with long, soft, yellowish or reddish fur (see photo B in the gallery). It occurs from northern Mexico to Brazil, and the other four species of the genus *Natalus* are restricted to either northern South America or the West Indies.

These bats' wing and hind-limb bones are long and spindly looking, and their ears are large and funnel shaped, as their common name indicates. These bats are commonly found roosting in caves and mine tunnels, often in fairly sizable aggregations. They are insectivorous, but details of their food habits are unknown. These curious small animals are but one of many such groups that have been little studied and are rarely encountered except by bat biologists and the occasional spelunker.

Thumbless Bats

In contrast to vampires, which have long, overdeveloped thumbs that are critical to their locomotion, the two species of bats in the family Furipteridae lack visible thumbs. Actually, the thumbs are there, but they are so tiny that they are invisible and are surely of little or no use to the bats. *Furipterus horrens*, also known as smoky bats in reference to their nondescript color pattern, are widely scattered through Central and South America but rarely encountered. In 30 years of fieldwork in the tropics, I have encountered only a single roost of this species.

That roost was in a jumble of human-size rocky boulders on an island in the middle of the Rio Xingu in the Brazilian Amazon. The bats were in the dark recesses formed by the boulders, which had been piled haphazardly by some former flood stage of the river. These tiny bats, each about the size of a modest butterfly, are fragile, spindly, and not unlike funnel-eared bats in overall delicate structure.

In fact, this family, along with the Natalidae and the Thyropteridae, to be discussed, may represent somewhat related relics of an old radiation of New World bats that are also related to the much more cosmopolitan Vespertilionidae. All are slightly built and similar in overall morphology, and all feed on small insects, which they capture in aerial pursuit. They are reported to fly like moths fairly low over the forest floor.

Sucker-Footed Bats

Myzopoda aurita is unique. Known only from the island of Madagascar, these tiny bats have large ears, a long tail that extends beyond the tail membrane, and small adhesive pads on their wrists and ankles. These suction disks are not as well developed as those in the family Thyropteridae (see next section), but they are likely quite useful in holding fast to smooth leaves.

Unfortunately, little is known of the natural history of this species. Their ears are unlike those of other long-eared species in that they are slender and unconnected. The tragus is also long but fused to the inner margin of the ear pinna itself. The antitragus, at the base of the ear conch, is modified into a padlike structure. Other long-eared bats in other parts of the world are frequently moth feeders that use low-intensity sounds of different frequencies. Recent work has shown that some of them cue in on sounds made by the prey animals themselves rather than rely solely on echolocation. Whether this practice also applies to *Myzopoda* is purely speculation at this point.

Disk-Winged Bats

New World disk-winged bats are similar to sucker-footed bats in that they have suction cups on their wrists and ankles. However, in the family Thyropteridae, the suction cups are much better developed, and the bats are capable of hanging from a pane of glass by a single wrist cup. There are three species of *Thyroptera,* one of which was named in 1994. It seems amazing that such an unusual creature would have gone undetected until so recently, but there are still many very poorly studied areas in the New World tropics, and undoubtedly more species of bats waiting to be discovered.

Spix's disk-winged bat (*Thyroptera tricolor*), the best known of the disk-winged bats, occurs widely through Central and South America (Figure 2.13). My mentor and colleague Jim Findley and I happened on a relatively dense population of these interesting little bats in the mid-1970s on the isolated Osa Peninsula of Costa Rica. We knew the bats had been reported to roost in the rolled leaves of bananas and their relatives, the heliconias. These leaves are newly developing and grow upward in tightly rolled tubes until their full length (an impressive 2.5 to 3.0 meters in some of the larger species) is extended, at which point they slowly unfurl. During the few days that it takes the leaves to open fully, they provide a hollow tube that serves as a shelter for these bats, as well as for a variety of other small creatures ranging from insects to frogs.

Once we discovered the right stage of leaf development, we quickly began to find small groups of the bats roosting in the leaves. We worked out a way to capture these groups, no small chore because the bats roost right side up (also exceptional in the bat world) and are ready to leap into flight at the slightest disturbance to the leaves. By marking the animals, we were able to recapture members of several groups for several days in succession and map their use of the habitat and its limited population of rolled leaves.

Each leaf contained a group consisting of one or more adult males, several adult females, and subadults of both sexes. Although the groups were unpredictable in sex and age composition, the overall sex ratio in the population was about even. An individual leaf lasts for only a day or two, so the groups have to shift their abodes constantly. These groups showed a definite cohesion; we rediscovered the same individuals together day after day.

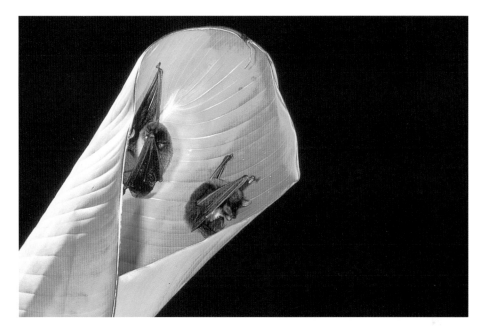

Figure 2.13. Spix's disk-winged bats, *Thyroptera tricolor,* emerging from their roost in an unfurling banana leaf in Panama. Suction cups on their ankles and wrists enable these insectivorous bats to climb with ease on slick leaf surfaces.

We estimated the roosting area of a single colony to be about 3,000 square meters and calculated a probable density of 3.3 colonies and about 20 bats per hectare. Upon completion of the study, we found that we had counted 3.7 colonies and 21.9 bats per hectare in our study area.

We carefully collected fecal samples from 21 leaves, and by calculating the average amount of fecal material per individual, we estimated that each bat consumes up to 0.8 grams of insects per night. This is a minimal estimate, because in-flight defecation may occur during feeding bouts. Details of their food habits remain unknown.

This species provided us with another example of direct communication between individuals. Starting with an entire group in a small cloth bag, we released them one by one after examining and marking them. One flew in decreasing circles around us and the bag and actually lighted on the bag several times. It eventually flew to a nearby rolled leaf of the appropriate size and, after circling a bit, disappeared inside. The next three bats we released went directly to that leaf, circled it, and joined their companion. The fifth bat went to a second rolled leaf nearby, and the next three went to the second leaf as well. The final bat went to the first leaf and joined his companions, although by this time the leaf was quite crowded. We had the definite impression that these animals were communicating by means of high-frequency calls beyond our hearing.

These bats are insectivorous and extremely adept at flying through the under-growth layer of the rain forest. They do not become entangled in mist nets, as do other bats, but instead simply land on the nets and sit quietly until disturbed.

Painted Bats

Kerivoula picta, the painted bat, is boldly colored. Its fur is bright orange or scarlet, and its wings are black, with orange extending along the fingers. Other species in this genus are also interestingly colored, ranging from reddish orange and black fur intermixed, with frosted tips on individual hairs, through reddish and yellowish species to more normally dull-colored ones.

This subfamily, Kerivoulinae, is one of several divisions of the family Vespertil-ionidae, one of the most common and widespread of all bat families. The single genus and 22 species of kerivouliines are also sometimes called woolly bats, because their fur is long and curly. They are small and delicate, and have large, pointed, fun-nel-shaped ears. They live in the forests of Asia and Africa, where they forage late in the evening by flying close to the ground. So far as we know all are insectivorous.

These bats roost in a variety of places, including buildings, foliage, tree holes, and similar sites. Some African species use the abandoned nests of weaver finches, a common resource in many areas. Strikingly colored painted bats have been re-ported roosting in a tree that retains variously orange and black dead leaves, which might well account for their otherwise conspicuous color pattern.

This is another poorly known group, with little known about their natural history. Scattered reports of reproduction suggest that a single young is born, but the species breed at different times in various parts of the two continents where they occur.

Little Brown Bats

If you live in the northern half of the United States, chances are pretty good that you have seen little brown bats (see Figure 1.22). *Myotis lucifugus,* a member of the family Vespertilionidae, is one of the most common species of bats foraging for in-sects on summer evenings throughout the United States. They have also adapted readily to human beings' tendency to build houses, barns, and other outbuildings that the bats assume were meant to be bat roosts. If you have a colony of bats in your attic, it is likely to be this species.

Little brown bats are small, brownish, and rather nondistinctive, even for bats. In the summer they roost in buildings of all types, caves, hollows, mines, and a va-riety of other natural and artificial refuges. In the winter they hibernate in cold caves in the northern part of their range. Their preferred feeding areas include rivers and small streams, where they feed on insect swarms either flying over the water or emerging from it. Their prowess at catching mosquito-size prey allows them to capture up to several hundred insects per hour.

These small bats have an interesting life cycle. They breed when the sexes come together at a limited number of hibernacula, or hibernating caves. However, it would be disadvantageous for the embryo to develop to maturity while the female is hibernating, because she would be unable to feed it when it was born. So the sperm are stored in the female's reproductive tract, and fertilization is delayed until she emerges from hibernation in the spring. This results in most young being born in June and July, when insect populations are reaching their peak throughout the bats' range. The females find plenty of food to withstand the extra energy demands of lactation, and the young grow rapidly, reaching sufficient size to begin flying at about three weeks of age.

The females band together to have their young, and the resultant "nursery colonies" may be fairly sizable, with groups of up to 1,000 not uncommon. These bats are fairly temperature tolerant, allowing them to use attics of buildings without heat distress. Their nursery colonies, and additional bachelor colonies comprising mostly males, break up in late summer and early fall, when both sexes begin to gather at the hibernacula. Sometimes these bats have to fly hundreds of kilometers to find a suitable cave or mine with stable, cold winter temperatures. They must spend several months each winter in hibernation, subsisting on the fat stores they have accumulated during the summer. Any disturbance to the colony during this time is likely to cause considerable damage. Bats that are aroused from torpor expend significant amounts of energy warming up, and repeated such disturbances might make it impossible for them to survive the winter.

These bats are frequently the cause of complaints from householders, because they tend to deposit guano in attics, and an occasional individual may stray into living quarters. However, they are superb insect controllers, and people who understand and appreciate them often are willing to put up with the inconvenience of sharing their buildings with little brown bats.

Tube-Nosed Insectivorous Bats

Tube-nosed insectivorous bats, *Murina*, are restricted to Asia, where the several species live in a variety of areas, although they seem to be more common in uplands. Their common name comes from the curious, tubelike extensions on these bats' nostrils. They are similar in this respect to tube-nosed fruit bats in the genera *Nyctimene* and *Paranyctimene*. This similarity has to be the result of convergent evolution, because the two groups are not closely related. The purpose of this curious structural adaptation remains obscure.

These bats are insectivorous, but little is known of their activities and haunts. They fly low over the surface of fields and grassy areas when foraging, but details of their food habits are unknown. They roost in caves and in the vegetation in dry leaves. There are reports of some species carrying twin fetuses, a somewhat uncommon occurrence in bats.

Murina are good examples of how little we know even about species that are widespread and not uncommon in many areas. These animals have curious anatomical adaptations, a somewhat uncertain evolutionary history, and ecological requirements that are almost totally unknown. There are examples of species like this on every continent, providing research topics for scores of scientists for years to come.

Long-Fingered Bats

Long-fingered bats, also known as long-winged or bent-winged bats, are another Old World group found widely distributed through Africa, Asia, and the Pacific islands (see photo J, *Miniopterus inflatus*, in the gallery). The second phalanx of the longest finger in their wings is elongated so much that the wings have to fold back on themselves when the bats are at rest. They fly rapidly but very erratically when foraging for small beetles and other insects at about 10 to 20 meters above the ground.

These bats hibernate in the northern parts of their range, and they show the characteristic reproductive adaptations to this behavior. However, unlike most vespertilionids, among which delayed fertilization is the norm, in *Miniopterus* implantation of the fertilized embryo is delayed. The result is the same, with birth delayed until environmental conditions are favorable, providing an abundant food supply for lactating females and subsequently for developing young. The length of the delay varies widely with latitude and apparently is determined by day length. Another curious difference in this group is that the male reproductive system regresses after mating; males of other species retain their ability to impregnate females even during hibernation.

Many of the species of *Miniopterus* form small colonies, although extremely large ones have also been reported. One study in South Africa reported over 100,000 juveniles in a single nursery cave. Like many other species these bats may be seasonally migratory, moving considerable distances between summer and winter roosts.

Short-Tailed Bats

Yet another strange bat is the New Zealand lesser short-tailed bat, *Mystacina tuberculata*. The only surviving member of the family Mystacinidae, this species is known only from New Zealand, and it is the only bat found in this small island country, save an occasional windblown stray from Australia and a widespread vespertilionid, *Chalinolobus tuberculatus*, the long-tailed wattled bat. Another species of mystacinid, *Mystacina robusta*, is thought to be extinct. Short-tailed bats are relatively small, with a very thick coat of fur. Their most distinguishing feature is an extra talon on the base of the large claws of the thumb and hind feet. No other bat has this feature.

Short-tailed bats are remarkable in that they are able to move freely along the branches and trunks of trees as well as on the ground. Their wings fold distinctively to allow for this increase in terrestrial and arboreal locomotion. They are also thought to use their incisors to make burrows, and they are the only species of bats known to roost in burrows.

Short-tailed bats also have one of the most catholic of bat diets. They apparently feed on insects as well as fruit, nectar, pollen, and perhaps even carrion. Their status as the only species of bats in New Zealand may have allowed them to expand their niche to encompass ecological space that is elsewhere occupied by other species.

These bats' reproductive pattern suggests delayed fertilization or implantation. They bear a single young in the middle of the austral summer, December. Copulations have been observed in May, which suggests a delay to account for the otherwise very long developmental time in comparison to other species of bats.

These bats are difficult to place phylogenetically. Their closest relatives are not obvious, although they are thought to be somewhat intermediate between the Vespertilionidae and the Molossidae. However recent molecular studies have suggested some relationship with the New World Phyllostomidae. Lack of a fossil record precludes the usual clues to classification.

Free-Tailed Bats
The family Molossidae, free-tailed bats, occurs worldwide in tropical and subtropical regions. If you have been to Carlsbad Caverns or other large caves in the southwestern United States, you will have seen the spectacular exit flight of Brazilian free-tailed bats, *Tadarida brasiliensis*. These animals form some of the largest colonies of vertebrates on earth, up to 20 million individuals in a single cave (see Figure 1.31).

Such large colonies play an important role in insect control over a large area of countryside. They also require constant, high population levels of insects. Therefore in the wintertime these bats migrate southward, where they form smaller colonies over a wide area.

Their combination of migratory habits and a penchant for insects put this species on a collision course with human beings a few years back. The colony at Carlsbad Caverns decreased steadily in size from several million to a couple of hundred thousand during the 1960s. In the early 1970s the National Park Service provided funding to determine the causes of this decline. We began a several-year study to document the decline and try to learn how to reverse it.

The first difficulty was in devising reliable, quantifiable census methods that would allow us to monitor the size of the population. Early estimates varied widely, but we felt secure that in the 1950s there had been several million bats in the colony. My colleagues Scott Altenbach and Ken Geluso of the University of New Mexico devised a clever photographic method that allowed us to take pictures of the exit flight every few seconds and then actually count the bats exiting over the course of the evening. This meant taking hundreds of photos during the several hours the exit flight might last, so we did not want to employ it every night!

Nevertheless, we were able to determine that the colony had indeed declined to just over 200,000 individuals, a potentially serious situation. We had strong clues from a variety of earlier studies that the culprit might be organochlorine hydrocarbons from heavy pesticide use in the Pecos Valley and other nearby foraging areas. The use of DDT was widespread in this country and in Mexico during the period of the decline, and the bats' insectivorous diet made them vulnerable to it. If a bat feeds on insects carrying small amounts of pesticide residues, it can quickly accumulate these small amounts into large ones. Vertebrates store the residues in their fat, and we knew that these bats built up large amounts of fat to prepare for their long migratory flight each fall.

Over three years we carefully documented these fat stores and at the same time analyzed numerous individuals of all ages to determine their pesticide load. The picture that emerged was not a pretty one. In brief, the bats were accumulating large amounts of pesticide residues, breakdown products of DDT, in their fatty tissues. This fat was then metabolized quickly during their fall migration, and we documented experimentally that flying bats could release enough pesticide to disrupt their physiology seriously, perhaps to kill them if sufficient quantities reached vulnerable brain tissue.

One other result was that female bats tended to unload their heavy pesticide burdens into their offspring through nursing. The fatty tissue in the mammary glands was also a storage site for the compounds, and the young of the year were the unfortunate recipients of this largesse. The subsequent inability of these young to cope with the burden and their failure to reach maturity were contributing to the decline in the bat population.

Fortunately the hazards of DDT were well documented by that time, and its use was outlawed in the United States. It continued to be used in Mexico, however, which meant that the animals were exposed during part of the year in many areas. Slowly but surely, though, the numbers of bats built back up in Carlsbad Caverns, and today it is once again possible to see absolutely spectacular exit and entrance flights there.

However, in 1991 Bat Conservation International conducted a survey of the major overwintering caves of this species and documented 95 to 100 percent losses in 5 of the 10 largest colonies and about 75 percent losses in 3 more. Only 2 caves held stable populations, and one of those has since been entirely destroyed. This species is unlikely ever to be as numerous as it once was, but it is still possible to see huge numbers of individuals at sites such as the Congress Avenue Bridge in Austin, Texas, where midsummer exit flights are quite impressive as the bats leave to forage up and down the river.

.3.

BATS AND HUMANS

DO BATS FLY INTO PEOPLE'S HAIR?

Although it is a prevalent myth, there is no evidence of bats flying into people's hair. The belief probably stems from common bat flight behaviors that are easily misinterpreted. Bats are naturally curious, and if you walk down a forest trail on a warm summer evening, it is not uncommon to have a bat fly very near to you before veering off. Bats become accustomed to flying certain pathways, and an unusual object such as a human being in a pathway may be a natural curiosity.

Another situation that might suggest this behavior occurs when many bats are disturbed in an enclosed area, such as a small cave roost. With lots of bats flying at once in a confined space, some may not only fly very close but occasionally even brush against people who get in the way. They are in fact trying to escape rather than to attack, and human hair holds no particular attraction for them. However, should a bat brush heavily against a particularly bouffant hairdo, it could conceivably become entangled and might well lead to an experience that gets handed down and expanded upon from generation to generation.

Brock Fenton, a Canadian bat biologist, has recently suggested two additional scenarios that might be origins for this myth. Once again imagine that summer evening, but this time recall a site where you have been surrounded by a swarm of small flying insects, such as mosquitoes or midges. It is certainly possible that such a congregation could attract a hungry bat, and in the process of feeding, the bat might come uncomfortably close to the human at the center of the swarm.

The other situation will be familiar if you have ever seen a bat trapped inside a room. The bat circles, obviously seeking an escape route, and as it approaches each wall, it slows almost to a stall, at which point it swoops low to regain speed and increase lift and thrust to continue the flight. These swoops can bring the animal very close to human occupants of the room and might easily be misinterpreted as attacks.

ARE BATS A THREAT TO HUMANS?

Bats are not a threat to humans in the sense of mounting a direct attack or flying into your hair. With the rare exception of vampire bats in Latin America, they will almost never approach a human with anything other than curiosity in mind. They can, however, harbor diseases that can affect humans.

DO BATS CARRY DISEASES?

The most serious disease bats can carry is rabies, and the potential threat of rabies argues against handling bats, particularly if they are acting strangely or let you approach without trying to escape. This said, it should be stressed that bats are no more likely to harbor rabies than many other mammals, and much less so than some species. In the wild only a fraction of a percent of bats are rabid. Also bats are much more likely to transmit rabies to other bats than to any other species of mammals. In most areas where rabies outbreaks occur, the strain of the virus has been traced back to dogs, cats, raccoons, skunks, or other mammals.

It is theoretically possible for any species of bat to become rabid. However, the likelihood is much greater for vampire bats than for others, simply because they come into direct contact with the blood of other mammals every day. Insectivorous species in northern latitudes also occasionally harbor the virus but never for very long, because they are also killed by the disease. Fruit eaters in the tropics almost never have rabies.

Another disease occasionally associated with bats is histoplasmosis, caused by the fungus *Histoplasma capsulatum*. It is possible for humans to become infected with this disease by breathing in the spores, which are sometimes found in large, dry, dusty accumulations of bird or bat droppings. Avoidance of these situations should virtually eliminate the danger of contracting this disease. If it is necessary for you to remove large accumulations of dry guano from an attic or other enclosed area, using a respirator should protect you from this risk.

Most species of bats also harbor ectoparasites such as mites and occasionally fleas or ticks. However, most such ectoparasites are highly host specific, meaning they live only on bats, and in fact frequently only on a single species of bat. Thus they pose no threat to humans. I have handled thousands of bats and been exposed to these ectoparasites many times over the years, with nary an ill effect. The most common of these are frequently bat flies, which may leave the bats and scamper across your hand or arm, but they do not bite and eventually leave of their own accord.

Diseased bats can theoretically be found anywhere. That is why bats that are out of place, acting strangely, or unafraid of humans should be avoided. However, the odds of you being killed by a bat-borne disease are much less than one in a million, and you face greater danger from honeybee stings or bathtub falls than from any sort of health risk from bats.

WHY DO BATS LIKE BUILDINGS?

Humans have created havoc with the environment of most other kinds of animals, but there are a few species that have adapted comfortably to the ways of people. Several species of bats have taken to buildings as if they own the places. Attics, barns, outbuildings, and the like mimic the ancient snag or cave roosts that were probably the ancestral refuges of these species. Such environments provide the requisite dark, relatively temperature-stable refuge from predators that most species of bats require for a daytime retreat. The fact that these quarters are shared with a family of *Homo sapiens* below is of little consequence to the bats. All they need is a tiny entrance and exit hole and freedom from disturbance.

WHAT SHOULD I DO IF A BAT GETS IN MY HOUSE?

The single most common bat problem that people are likely to face is the appearance of a bat inside the house. This can be startling at first. However, there is little to fear, and easy remedies are at hand. The bat is likely to be as confused and upset as you are. It is undoubtedly lost and trying to make its way outside, so all you need to do is help it along by opening doors and windows and letting it find a way out. If this fails you can gently put a towel over the animal when it has landed on the wall or floor, then carry it outside to release it.

HOW CAN BAT PROBLEMS BE AVOIDED?

The other common bat-related problem is a colony using the attic of your home. The only solution here is to build the unwelcome visitors out by finding and blocking all potential entrances with bird netting. Of course this needs to be done when the bats are not present. The best time to do repairs is during the winter, when the bats will likely have moved to cooler quarters to hibernate.

If your bats are year-round residents, you will need to do the repairs at night, when they are out foraging. This will also require ensuring that there are no young left behind so that you do not inadvertently trap them inside. In North America, June, July, and August are the most likely times for young to be in the roost.

ARE THERE ANY EFFECTIVE BAT REPELLENTS?

Should you have bats that need to be discouraged from roosting in an inconvenient location, there is no easy solution. Although bats in your attic probably pose little danger, and may in fact be a considerable benefit in control of noxious insects, sometimes the accumulation of guano or the presence of odors makes it desirable to discourage them. The only effective means of doing this is to build them out, closing all entrances and exits and plugging all small holes that might allow access to the space (see *How Can Bat Problems Be Avoided?*).

To date all electronic and chemical means have been ineffective in repelling bats. This includes all the common chemicals commercial pest-control firms will offer to pump into your location. Such a cure is likely to be far more harmful to you than to the bats. Ultrasonic devices designed to interfere with bats' echolocation system have proved equally ineffective.

Lights or fans in a few particular situations have had limited success. These relatively innocuous methods can be tried with little likelihood of harmful side effects, but their chances of long-term success may not be great. Similarly, short-term results may be possible with commercial spray products designed to keep pets off furniture. Also, exclusion is often relatively easy, safe, and permanent.

HOW CAN I ATTRACT BATS?

Many people now recognize the importance of maintaining active bat colonies nearby and are constructing a variety of artificial roosts in order to attract bats (Figure 3.1). Bat houses are becoming increasingly popular and may provide an excellent way to attract bats without having to share your own house with them. Although bat houses have been used in Europe for decades, they have only recently become common in the United States, and little is known about how to maximize their effectiveness (Figure 3.2).

One valuable contribution available from Bat Conservation International is *The Bat House Builder's Handbook*, by Merlin D. Tuttle and Donna L. Hensley. This booklet contains a wealth of background on the importance and value of bat houses as well as practical information on how to build them, where to site them, and other secrets to successful occupation of bat houses by various species.

Figure 3.1. A bat house mounted on a pole near Safford, Arizona.

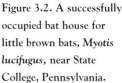

Figure 3.2. A successfully occupied bat house for little brown bats, *Myotis lucifugus*, near State College, Pennsylvania.

Bat Conservation International is currently conducting a research project on bat house use with volunteers in 49 states, several Canadian provinces, and some additional countries. Volunteers gather data on standardized forms to report to the central database in Austin, Texas. Anyone wishing to participate in this project can write for further information to North American Bat House Research Project, Bat Conservation International, P.O. Box 162603, Austin, Texas 78716.

WHAT SHOULD I DO IF I FIND A SICK BAT?

A sick bat should be left strictly alone. Call your local animal control office to deal with the animal. If you have to remove such an animal from the house, use thick gloves and make sure the bat is placed where it will not be found by pets or other animals. Bats that are flying normally are unlikely to be sick, but animals that are on the ground or unable to fly should be assumed to be sick and treated with caution.

WHY DO PEOPLE DISLIKE BATS?

It is probably unfair to say that most people dislike bats nowadays. Although bats certainly have suffered from myths, legends, and bad press over the years, the negative attitudes toward them seem to be turning around recently. As people become better educated about bats, their attitudes change from either indifference or actual dislike to interest, wonder, and even appreciation.

The early associations of bats with darkness and evil were certainly unwarranted. In addition, the occasional incidence of rabies in bats led to a sense of distrust that has also proved unfounded. As educational programs continue and our knowledge of the valuable services bats provide all over the world increases, the outlook for their protection and continued contributions is improving.

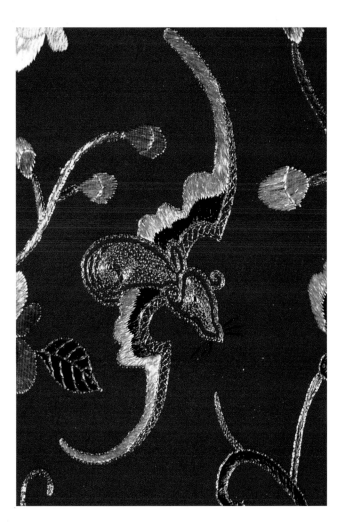

Figure 3.3. A stylized bat on a red mat at a Q'ing dynasty Buddhist shrine.

WHAT DIFFERENT CULTURAL ATTITUDES ARE THERE ABOUT BATS?

Certainly bats are not disliked or mistrusted in all cultures. In China, for instance, bats are considered good luck, and a stylized bat symbol is used as a good luck charm (Figure 3.3). In North America, Navajo culture considers bats mentors of the night and identifies them with a prominent deity, Talking God.

Playing a prominent role in a culture does not, however, always ensure ideal treatment for bats. The Chamorro people, native to the island of Guam and the Northern Mariannas, consider fruit bats important to their culture and celebrate that link by eating them as ceremonial dishes on traditional feast days. As a result fruit bats have become very scarce on those islands; they have even been imported from other island groups to satisfy the culturally based market.

WHAT ARE SOME MYTHS AND LEGENDS ABOUT BATS?

Gary McCracken, a professor at the University of Tennessee who studies bats, has assembled a considerable amount of information on them for *The Encyclopedia of American Folklore and Superstition*, and he also summarized that information in a series of six articles in *Bats*, the quarterly publication of Bat Conservation International. He concludes that bats are frequently seen as liminal animals, which means that they do not fit into people's view of the normal scheme of things. They tend to be in between and thus difficult for people to identify with. For instance, people frequently are not sure whether bats are mammals or birds.

North American Indian tribes have a variety of myths and legends about bats. One has bats being created by birds to help them win a ball game with the animals. This legend also accounts for the origin of the flying squirrel in much the same fashion.

There are variations on the ambivalent quality of bats from many cultures. On the Pacific islands of Samoa and Fiji, legends have bats originating by stealing the wings of birds. One variant of this legend reverses the roles: flying foxes originally walked on all fours and rats had wings. The bats borrowed the rats' wings and never returned them, causing lasting enmity, as one might well imagine. This is the reason rats climb trees and try to eat young bats.

One especially engaging legend comes from the Pomo Indians of California. Their myth suggests a bat that can eat obsidian and spit out arrowheads. In fact, there is a bat called the California leaf-nosed bat, *Macrotus californicus*, that has an arrowhead-shaped nose leaf (see photo F in the gallery). Scholars speculate that this distinctive-looking feature led to the legend of the arrowhead-producing bat.

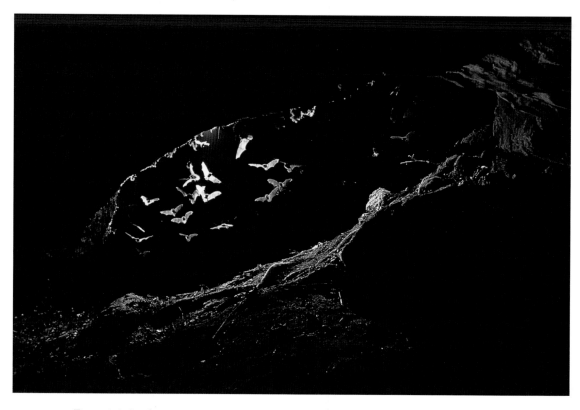

Figure 3.4. Southeastern myotis, *Myotis austroriparius*, and endangered gray bats, *Myotis grisescens*, returning at dawn to their nursery roost in Judges Cave, Florida. This cave is now a protected bat sanctuary.

WHAT GOOD ARE BATS?

This is frequently the first question asked when one tries to make the case for bat conservation. Fortunately, it is also an easy question to answer because the value of bats is obvious once one understands a bit about their biology.

For temperate zone bats, the main value is insect control. These little insectivores put away enormous quantities of insects every summer night, all over the northern and southern hemispheres (Figure 3.4). Having a bat house instead of an electronic bug zapper in your backyard makes for far more pleasant cookouts. Bats' efforts at pest control are also a boon for orchardists and farmers. In addition, the guano that bats produce is a valuable fertilizer, in both natural ecosystems and agricultural systems (Figure 3.5).

These insectivores have the added benefit of serving as indicators of the health of the ecosystem. The free-tailed bat declines of the 1960s were a wake-up call to

Figure 3.5. A cave in Thailand that contains approximately 500,000 wrinkle-lipped free-tailed bats, *Chaerephon plicata*, which are important sources of guano for fertilizer. Annual sales of guano from single caves amount to more than $50,000 (US) and often are a major source of income for local villages. Huge bat colonies such as this consume many thousands of kilograms of insects nightly. Bats in nearby caves are now largely extinct as a result of overharvesting for human food. To protect these bats, the owner pays a guard to spend each night at the cave.

the dangers of long-lived pesticides in the environment. Such pollution signals are going to be increasingly important in our world of ever-expanding human populations and economic development.

In the tropics bats are even more valuable to the functioning of natural and agricultural environments (Figure 3.6). They serve as pollinators and seed dispersal agents for hundreds, and probably thousands, of species of tropical plants (Figure 3.7). Many plants have adapted to the benefits of this service by opening flowers and producing nectar only at night to attract the bats (Figure 3.8). Others produce fruits whose seeds are distributed well away from the parent plants by flying bats (Figure 3.9).

These ecosystem services are vitally important to virtually every major habitat type

Figure 3.6. (left) A short-nosed fruit bat, *Cynopterus sphinx*, eagerly licking the nectar reward from a wild banana flower. The bat's face will soon become covered with pollen, which it will carry to the next plant, thereby ensuring that the plant will bear fruit and continue the cycle. Even though cultivated bananas are vegetatively reproduced on plantations, bat pollination is critical to the survival of genetic strains of wild progenitors that could someday prove essential to the development of new, more productive, or disease-resistant varieties. **Figure 3.7. (right) A big fruit bat, *Artibeus lituratus*, pollinating a flower of the neotropical tree *Pseudobombax*.** This flower opens at dusk and falls off by morning. Like many bat-dependent flowers, it is white, enabling bats to see it more easily on dark nights.

A.

B.

Figure 3.8. A Wahlberg's epauletted fruit bat, *Epomophorus wahlbergi*, (A) coming to a baobab flower to feed, (B) its body collecting and distributing pollen in the process. Like many plants that depend on bats for reproduction, the baobab has blossoms that are lightly colored and scented, and thus easier for nocturnal feeders to find.

Figure 3.9. A southern long-nosed bat, *Leptonycteris curasoae*, approaching a cardon cactus fruit. Eating in flight, the bat repays the cactus for the luscious midnight snack by dispersing its seeds. In the spring, as lesser long-nosed bats begin their northward migration from southern Mexico, they feed on the nectar from flowers of the saguaro, organ-pipe, and cardon cacti. During many nights of feeding, the cacti are inadvertently pollinated as the bats probe the flowers for their sweet treat. Later in the season the pollinated flowers develop into rich fruits, laden with thousands of seeds. As more bats pass through the area in June, they perform another service by eating the ripe fruits and dispersing the seeds. This bat species is thus essential to some of the Sonoran Desert's most important plants, and the fact that the bats' numbers appear to be declining raises concern for the entire ecosystem.

on earth. Bats are missing only from polar ecosystems and a few small, remote oceanic islands. The loss of significant numbers of species of bats would have far-reaching consequences for the other animals and the plants that share their communities.

Recent advances in pharmacological studies have suggested potentially valuable uses for the anticoagulant compounds found in vampire bats' saliva. Prospecting for new drugs is an important growth industry in many parts of the developing world, and finding such a potentially valuable resource in an animal that has been severely persecuted as a pest is an interesting object lesson in why we should protect all our natural resources.

Beyond all these practical reasons for protecting a valuable resource, bats are part of our natural heritage, to be enjoyed for purely aesthetic reasons. Furthermore, surely these remarkable products of millions of years of evolution have every right to exist for their own sake, regardless of any value system imposed by humans.

HOW CAN I BECOME A BAT BIOLOGIST?

Studying bats is a rewarding career, especially for people with an abiding curiosity about the natural world and its inhabitants. Bat biologists enter the field from a variety of disciplines, but the majority are mammalogists. Mammalogy is the discipline of zoology devoted to studying mammals, and bats are one of the most diverse and poorly known groups of mammals.

Other bat biologists are interested in particular systems; for instance, neurobiologists are frequently interested in echolocation. Echolocation is uncommon among animals, and the principles learned from the study of bats are invaluable to studies of the physics of sound transmission and the processing of sounds in animal brains. Physiologists interested in the functioning of various systems in animals and in humans study bats as unique examples of flying mammals and of animals that spend a considerable part of their lives upside down, requiring interesting adaptations in the circulatory system, among other things.

However one reaches the goal of studying bats as a profession, the first step is to obtain a good undergraduate education in biology. Although it is not possible to specialize in mammalogy as an undergraduate, it is possible to take courses in mammalogy, field biology, and at some universities even special seminars in bat biology. Courses in ecology, behavior, botany, and entomology will also provide good background. Other biological subjects that will be useful include genetics, physiology, comparative anatomy, systematics, and techniques of molecular biology. Depending on your future specialty, additional courses in foreign languages (much of the scientific literature is in languages such as German, French, Russian, Japanese, and Chinese, and fieldwork in Latin America requires Spanish and Portuguese), statistics and other areas of mathematics, and computer techniques will be very useful.

The next step is to choose a graduate school. Contacts made through your undergraduate professors will assist in this process. If you know that you are going to specialize in mammalogy, ecology, or behavior, you should select a school with a mammalogist, vertebrate zoologist, or behaviorist on staff. If you definitely intend to work on bats, it makes sense to choose someone who also specializes in them. This will ensure not only that you get the proper mentoring from your major professor but that you will be part of a cadre of graduate students with common interests to assist in your education.

There are many universities with good graduate programs in mammalogy, ecology, or behavior, both in the United States and abroad. Additional information on careers in mammalogy can be obtained by contacting the American Society of Mammalogists, the major professional organization for people in this discipline. The society welcomes student members, and information on becoming a member can be obtained by writing the secretary-treasurer, listed on the inside covers of issues of the *Journal of Mammalogy*.

Figure 3.10. A bat biologist showing a red bat, *Lasiurus borealis,* at a Bat Conservation International workshop in Pennsylvania.

With a little diligence you can also discover if there is a bat biologist in your area. Looking at articles in the *Journal of Mammalogy,* or at the bibliographies of this and other books and articles about bats, will give you names of bat biologists and some idea of their professional interests. Volunteering your services at a local university, museum, or zoo where bat studies are going on can be a rewarding way to enter the field and gain valuable experience.

Today you can find bat biologists employed by a wide variety of governmental and private organizations. Universities provide a major source of employment, but many bat biologists also work for museums and zoological parks. In addition, federal agencies—such as the U.S. Fish and Wildlife Service, the National Park Service, the Forest Service, and the U.S. Geological Survey—employ bat biologists. Also many private environmental organizations and consulting firms offer employment opportunities (Figure 3.10).

HOW DO SCIENTISTS CATCH BATS?

Bat biologists use a variety of methods to catch bats in order to study them. We actually knew very little about bat natural history until about a half century ago, when Japanese mist nets came into use for capturing bats. These very fine nets, almost like hair nets, are about 2 meters high and range from about 5 to 30 meters

Figure 3.11. (left) A spotted bat, *Euderma maculatum,* entangled in a mist net. Figure 3.12. (right) A bat biologist removing a big brown bat, *Eptesicus fuscus,* from a mist net.

long. They can be strung across flyways, such as paths in a forest, or over water sources, such as ponds and creeks, to catch bats as they fly along. The bats hit the nets and become entangled in them without suffering injury (Figure 3.11). They can then be removed carefully and either marked and released for later recapture or taken to the laboratory for further study (Figure 3.12).

Variations on this theme have been developed more recently. Denny Constantine, a veterinary biologist who has contributed greatly to our knowledge of bats and diseases, developed a type of funnel trap to use in the openings of large caves; it catches the bats as they exit (Figure 3.13). Merlin Tuttle refined this device, and his smaller version has become known as the Tuttle trap. Basically, it consists of two parallel rows of monofilament line strung vertically between metal frames. The bats hit the lines and slide down into a plastic catch bag, where plastic flaps keep them from escaping. These traps are portable and easy to hoist into cave entrances or into the forest canopy as needed.

WHAT IS LEFT TO LEARN ABOUT BATS?

Bat biology has come a long way in the last couple of centuries and has grown very rapidly in the last couple of decades. We now understand much more about the biology of bats in all areas than we did in the 1960s, when my own interest in bats began. The advances in understanding sophisticated systems such as echolocation have been particularly noteworthy.

Nevertheless, we have a considerable amount to learn about this important and interesting group of animals. Much of our knowledge about bat biology comes from a relatively few species (Figure 3.14). With over 925 species to choose from, there are still many bats that are almost completely unknown (Figure 3.15). Even the ba-

Figure 3.13. Constantine and Tuttle traps can be used to catch gray bats (*Myotis grisescens*) as they emerge from this cave in Tennessee.

Figure 3.14. The southern long-nosed bat, *Leptonycteris curasoae*, in flight to a cardon cactus flower. This is one species that has become much better known recently, thanks to several years of hard work by a dedicated team of bat biologists.

Figure 3.15. An Underwood's long-tongued bat, *Hylonycteris underwoodi,* a relatively common yet quite poorly known species of nectar-feeding bat widely distributed in Latin America.

sic classification of bats is little more than the most recent version of a phylogenetic hypothesis, which undergoes adjustment with each new study that is published.

Our knowledge of the interactions of bats with their environment is of critical importance to conservation biology (Figure 3.16), an area that is going to grow considerably in the early twenty-first century. Students interested in bat ecology and behavior will have ample opportunity to contribute to the field (Figure 3.17), especially in tropical regions, where the bulk of bat diversity is concentrated and where we understand the least about the world's most complicated ecosystems (Figure 3.18).

WHERE CAN I GET MORE INFORMATION ABOUT BATS?

There are several sources of information on all these topics. The references in the bibliographies contain a wealth of information on every aspect of bat biology. Your local library should have at least a sampling of them.

There are also organizations devoted to correcting misinformation about bats. The oldest and largest of these is Bat Conservation International, in Austin, Texas. This is a nonprofit membership organization formed to document and publicize the value and conservation needs of bats, to promote conservation and research pro-

Figure 3.16. A Commissaris's nectar bat, *Glossophaga commissarisi*, another widespread but poorly known species, whose interactions with the environment play an important but as yet largely undocumented role in the functioning of tropical ecosystems.

Figure 3.17. A black flying fox, *Pteropus alecto*, seen here pollinating a bottle brush flower. This bat is an example from the Old World of a worthy subject for the study of bat ecology.

Figure 3.18. White fruit bats, *Ectophylla alba*. Spectacular tropical bats such as these will attract the attention of future students of bat biology.

jects, and to assist with management objectives worldwide. For additional information write: Bat Conservation International, P.O. Box 162603, Austin, Texas 78716.

Some bat conservation groups in other countries include the following:

Australia

Australasian Bat Society
Lawrie Conole, Newsletter Editor
2/45 Virginia Street
Newton 3220, Victoria, Australia
phone 03-9-669-9732
fax 9-663-3669
e-mail (lconole@mov.vic.gov.au)

Canada

Bat Conservation Society of Canada
P.O. Box 56042, Airways Postal Outlet
Calgary, Alberta T2E 8K5, Canada
phone 403-860-BATS
e-mail (BCSC@cadivision.com)

England

The Bat Conservation Trust
Colin Catto, Scientific Officer
15 Cloisters House
8 Battersea Park Road
London SW8 4BG, England
phone 0171-627-2629
fax 0171-627-2628
e-mail (batcontrust@gn.apc.org)

Germany

AG-Fledermausschutz BaWue
Universität Tübingen
Auf der Morgenstelle 28
72076 Tübingen, Germany

Koordinationsgruppe
Fledermausschutz in Deutschland
K. Richarz, Chairman
Staatliche Vogelschutzwarte
Steinauer Strasse 44
D-60386 Frankfurt, Germany
phone 0049-69-411532
fax 0049-69-425152

Naturschutzbund Deutschland
BAG Fledermausschutz
Dr. J. Haensel
Hauptstrasse 13
D-13055 Berlin, Germany

Greece

Hellenic Bat Society
P.O. Box 3277
Gr-102 10 Athens, Greece
phone/fax 301-330-2548

The Netherlands

Vleermuiswerkgroep Nederland, VLEN
Emmalaan 41
NL 3581 HP Utrecht, The Netherlands
phone/fax 030-2544642

South Africa

Bat Interest Group
c/o Dr. Peter Taylor
Durban Natural Science Museum
P.O. Box 4085
Durban 4000, South Africa
phone 31-300-6218
fax 31-300-6302

Switzerland

Bat Conservation Switzerland
Fledermausschutz SSF/KOF
Zoological Museum of the University
Winterthurerstrasse 190
CH-8057 Zurich, Switzerland
phone 01-257-4776
fax 01-364-0295
e-mail (nyct240@zoolmus.unizh.ch)

Centre de coordination ouest
pour l'étude et la protection des
chauves-souris
Muséum de l'histoire naturelle
Case postale 6434
CH-1211 Geneva 6, Switzerland
phone 022-735-9130
fax 022-735-3445
e-mail (moeschl@mail.ville-ge.ch)

In addition, there is an e-mail discussion list called "Batline," which can be reached at batline@unmvma.unm.edu.

APPENDIX

CONSERVATION STATUS OF BATS OF THE WORLD (ORDER CHIROPTERA)

Scientific names are based on K. F. Koopman in D. E. Wilson and D. M. Reeder, eds., *Mammal Species of the World* (Washington, D.C.: Smithsonian Institution Press, 1993). Conservation status was derived from the literature and from an extensive questionnaire distributed worldwide to mammalogists; categories follow International Union for Conservation of Nature and Natural Resources Red Data Books. Common names are listed for genera and families, except where there is only a single species in the genus or family. In those cases the common name for the species is the common name for the genus or family as well.

Scientific Name	Common Name	Conservation Status
Family Pteropodidae—Old World fruit bats		
Genus *Acerodon*	Island fruit bats	
A. celebensis	Sulawesi fruit bat	No assessment
A. humilis	Talaud fruit bat	No assessment
A. jubatus	Golden-capped fruit bat	Endangered
A. leucotis	Palawan fruit bat	No assessment
A. lucifer	Panay giant fruit bat	Extinct >50 years
A. mackloti	Sunda fruit bat	Stable
Genus *Aethalops*		
A. alecto	Pygmy fruit bat	Stable
Genus *Alionycteris*		
A. paucidentata	Mindanao pygmy fruit bat	Potentially vulnerable
Genus *Aproteles*		
A. bulmerae	Bulmer's fruit bat	Endangered
Genus *Balionycteris*		
B. maculata	Spotted-winged fruit bat	Stable
Genus *Boneia*		
B. bidens	Manado fruit bat	Potentially vulnerable
Genus *Casinycteris*		
C. argynnis	Short-palated fruit bat	Potentially vulnerable
Genus *Chironax*		
C. melanocephalus	Black-capped fruit bat	Stable
Genus *Cynopterus*	Short-nosed fruit bats	
C. brachyotis	Malaysian short-nosed fruit bat	Stable
C. horsfieldi	Horsfield's short-nosed fruit bat	Stable

Scientific Name	Common Name	Conservation Status
C. nusatenggara	Nusatenggara short-nosed fruit bat	No assessment
C. sphinx	Short-nosed fruit bat	Stable
C. titthaecheileus	Indonesian short-nosed fruit bat	Stable
Genus Dobsonia	Naked-backed fruit bats	
D. beauforti	Beaufort's naked-backed fruit bat	No assessment
D. chapmani	Negros naked-backed fruit bat	Probably extinct
D. emersa	Biak naked-backed fruit bat	No assessment
D. exoleta	Sulawesi naked-backed fruit bat	No assessment
D. inermis	Solomons naked-backed fruit bat	Stable
D. minor	Lesser naked-backed fruit bat	Stable
D. moluccensis	Moluccan naked-backed fruit bat	Stable
D. pannietensis	Panniet naked-backed fruit bat	No assessment
D. peroni	Western naked-backed fruit bat	Potentially vulnerable
D. praedatrix	New Britain naked-backed fruit bat	No assessment
D. viridis	Greenish naked-backed fruit bat	Stable
Genus Dyacopterus		
D. spadiceus	Dyak fruit bat	Potentially vulnerable
Genus Eidolon	Eidolon fruit bats	
E. dupreanum	Madagascar fruit bat	Stable
E. helvum	Straw-colored fruit bat	Stable
Genus Epomophorus	Epauletted fruit bats	
E. angolensis	Angolan epauletted fruit bat	Stable
E. gambianus	Gambian epauletted fruit bat	Stable
E. grandis	Lesser Angolan epauletted fruit bat	Stable
E. labiatus	Ethiopian epauletted fruit bat	Stable
E. minimus	East African epauletted fruit bat	No assessment
E. wahlbergi	Wahlberg's epauletted fruit bat	Stable
Genus Epomops	Epauletted bats	
E. buettikoferi	Buettikofer's epauletted bat	Vulnerable
E. dobsoni	Dobson's epauletted bat	Stable
E. franqueti	Franquet's epauletted bat	Stable
Genus Haplonycteris		
H. fischeri	Fischer's pygmy fruit bat	Vulnerable
Genus Harpyionycteris	Harpy fruit bats	
H. celebensis	Sulawesi harpy fruit bat	Vulnerable
H. whiteheadi	Whitehead's harpy fruit bat	Stable
Genus Hypsignathus		
H. monstrosus	Hammer-headed fruit bat	Stable
Genus Latidens		
L. salimalii	Salim Ali's fruit bat	Stable
Genus Megaerops	Tailless fruit bats	
M. ecaudatus	Temminck's tailless fruit bat	Stable
M. kusnotoi	Javan tailless fruit bat	Potentially vulnerable
M. niphanae	Niphan's tailless fruit bat	No assessment
M. wetmorei	White-collared tailless fruit bat	Stable
Genus Micropteropus	Dwarf epauletted fruit bats	
M. intermedius	Hayman's dwarf epauletted fruit bat	Stable
M. pusillus	Peters's dwarf epauletted fruit bat	Stable
Genus Myonycteris	Little collared fruit bats	
M. brachycephala	São Tomé little collared fruit bat	Vulnerable
M. relicta	East African little collared fruit bat	Vulnerable
M. torquata	Little collared fruit bat	Stable
Genus Nanonycteris		
N. veldkampi	Veldkamp's dwarf epauletted bat	Stable
Genus Neopteryx		
N. frosti	Small-toothed fruit bat	Potentially vulnerable

Continued

Scientific Name	Common Name	Conservation Status
Family Pteropodidae—Old World fruit bats (*continued*)		
Genus *Nyctimene*	Tube-nosed fruit bats	
N. *aello*	Broad-striped tube-nosed bat	Stable
N. *albiventer*	Common tube-nosed bat	Stable
N. *celaeno*	Dark tube-nosed bat	No assessment
N. *cephalotes*	Pallas's tube-nosed bat	Stable
N. *certans*	Mountain tube-nosed bat	Stable
N. *cyclotis*	New Guinea tube-nosed bat	Stable
N. *draconilla*	Dragon tube-nosed bat	Stable
N. *major*	Island tube-nosed bat	Stable
N. *malaitensis*	Malaita tube-nosed bat	No assessment
N. *masalai*	Demonic tube-nosed bat	No assessment
N. *minutus*	Lesser tube-nosed bat	Stable
N. *rabori*	Philippine tube-nosed bat	Endangered
N. *robinsoni*	Queensland tube-nosed bat	Stable
N. *sanctacrucis*	Nendo tube-nosed bat	Probably extinct
N. *vizcaccia*	Umboi tube-nosed bat	Stable
Genus *Otopteropus*		
O. *cartilagonodus*	Luzon fruit bat	Potentially vulnerable
Genus *Paranyctimene*		
P. *raptor*	Unstriped tube-nosed bat	Stable
Genus *Penthetor*		
P. *lucasi*	Lucas's short-nosed fruit bat	Stable
Genus *Plerotes*		
P. *anchietai*	D'Anchieta's fruit bat	Stable
Genus *Ptenochirus*	Dog-faced fruit bats	
P. *jagori*	Jagor's dog-faced fruit bat	Stable
P. *minor*	Lesser dog-faced fruit bat	Stable
Genus *Pteralopex*	Monkey-faced bats	
P. *acrodonta*	Fijian monkey-faced bat	Endangered
P. *anceps*	Bougainville monkey-faced bat	Endangered
P. *atrata*	Guadalcanal monkey-faced bat	Endangered
P. *pulchra*	Montane monkey-faced bat	No assessment
Genus *Pteropus*	Flying foxes	
P. *admiralitatum*	Admiralty flying fox	Stable
P. *aldabrensis*	Aldabra flying fox	Vulnerable
P. *alecto*	Black flying fox	Stable
P. *anetianus*	Vanuatu flying fox	Vulnerable
P. *argentatus*	Ambon flying fox	No assessment
P. *brunneus*	Brown flying fox	Probably extinct
P. *caniceps*	North Moluccan flying fox	Stable
P. *chrysoproctus*	Moluccan flying fox	No assessment
P. *conspicillatus*	Spectacled flying fox	Stable
P. *dasymallus*	Ryukyu flying fox	Endangered
P. *faunulus*	Erabu flying fox	No assessment
P. *fundatus*	Banks flying fox	No assessment
P. *giganteus*	Indian flying fox	Stable
P. *gilliardi*	Gilliard's flying fox	No assessment
P. *griseus*	Gray flying fox	Stable
P. *howensis*	Ontong Java flying fox	No assessment
P. *hypomelanus*	Variable flying fox	Stable
P. *insularis*	Chuuk flying fox	Endangered
P. *leucopterus*	Philippine white-winged flying fox	Vulnerable
P. *livingstonei*	Comoro black flying fox	Endangered
P. *lombocensis*	Lombok flying fox	Stable
P. *lylei*	Lyle's flying fox	Stable

Scientific Name	Common Name	Conservation Status
P. macrotis	Big-eared flying fox	Stable
P. mahaganus	Sanborn's flying fox	Vulnerable
P. mariannus	Marianas flying fox	Endangered
P. mearnsi	Mearns's flying fox	No assessment
P. melanopogon	Black-bearded flying fox	Stable
P. melanotus	Black-eared flying fox	Stable
P. molossinus	Caroline flying fox	Endangered
P. neohibernicus	Great flying fox	Stable
P. niger	Mascarene flying fox	Vulnerable
P. nitendiensis	Temotu flying fox	No assessment
P. ocularis	Seram flying fox	No assessment
P. ornatus	Ornate flying fox	Potentially vulnerable
P. personatus	Masked flying fox	No assessment
P. phaeocephalus	Mortlock flying fox	Endangered
P. pilosus	Palau flying fox	Extinct >50 years
P. pohlei	Geelvink Bay flying fox	No assessment
P. poliocephalus	Gray-headed flying fox	Stable
P. pselaphon	Bonin flying fox	Vulnerable
P. pumilus	Golden-mantled flying fox	Vulnerable
P. rayneri	Solomons flying fox	Stable
P. rodricensis	Rodrigues flying fox	Endangered
P. rufus	Madagascar flying fox	Stable
P. samoensis	Samoan flying fox	Endangered
P. sanctacrucis	Santa Cruz flying fox	No assessment
P. scapulatus	Red flying fox	Stable
P. seychellensis	Seychelles flying fox	Stable
P. speciosus	Philippine flying fox	Stable
P. subniger	Dark flying fox	Extinct >50 years
P. temmincki	Temminck's flying fox	No assessment
P. tokudae	Guam flying fox	Probably extinct
P. tonganus	Pacific flying fox	Stable
P. tuberculatus	Vanikoro flying fox	No assessment
P. vampyrus	Red-necked flying fox	Stable
P. vetulus	New Caledonia flying fox	Stable
P. voeltzkowi	Pemba flying fox	Endangered
P. woodfordi	Dwarf flying fox	No assessment
Genus Rousettus	Rousette fruit bats	
R. aegyptiacus	Egyptian rousette	Stable
R. amplexicaudatus	Geoffroy's rousette	Stable
R. angolensis	Angola rousette	Stable
R. celebensis	Sulawesi rousette	Stable
R. lanosus	Long-haired rousette	Stable
R. leschenaulti	Leschenault's rousette	Stable
R. madagascariensis	Madagascar rousette	Stable
R. obliviosus	Comoro rousette	Stable
R. spinalatus	Bare-backed rousette	No assessment
Genus Scotonycteris	West African fruit bats	
S. ophiodon	Pohle's fruit bat	Stable
S. zenkeri	Zenker's fruit bat	Stable
Genus Sphaerias		
S. blanfordi	Blanford's fruit bat	No assessment
Genus Styloctenium		
S. wallacei	Stripe-faced fruit bat	No assessment
Genus Thoopterus		
T. nigrescens	Swift fruit bat	No assessment
Genus Eonycteris	Dawn fruit bats	
E. major	Dulit fruit bat	Vulnerable

Continued

Scientific Name	Common Name	Conservation Status
Family Pteropodidae—Old World fruit bats *(continued)*		
E. spelaea	Long-tongued dawn fruit bat	Vulnerable
Genus *Macroglossus*	Long-tongued fruit bats	
M. minimus	Small long-tongued fruit bat	Stable
M. sobrinus	Hill long-tongued fruit bat	Stable
Genus *Megaloglossus*		
M. woermanni	Woermann's bat	Stable
Genus *Melonycteris*	Black-bellied fruit bats	
M. aurantius	Orange fruit bat	No assessment
M. melanops	Black-bellied fruit bat	Stable
M. woodfordi	Woodford's fruit bat	Stable
Genus *Notopteris*		
N. macdonaldi	Long-tailed fruit bat	Stable
Genus *Syconycteris*	Blossom bats	
S. australis	Southern blossom bat	Stable
S. carolinae	Halmahera blossom bat	No assessment
S. hobbit	Moss-forest blossom bat	Stable
Family Rhinopomatidae		
Genus *Rhinopoma*	Mouse-tailed bats	
R. hardwickei	Hardwicke's mouse-tailed bat	No assessment
R. microphyllum	Large mouse-tailed bat	Potentially vulnerable
R. muscatellum	Small mouse-tailed bat	No assessment
Family Craseonycteridae		
Genus *Craseonycteris*		
C. thonglongyai	Hog-nosed bat	Endangered
Family Emballonuridae—sheath-tailed, sac-winged, and ghost bats		
Genus *Balantiopteryx*	Least sac-winged bats	
B. infusca	Ecuadorian sac-winged bat	Potentially vulnerable
B. io	Thomas's sac-winged bat	Potentially vulnerable
B. plicata	Peters's sac-winged bat	Stable
Genus *Centronycteris*		
C. maximiliani	Shaggy bat	Stable
Genus *Coleura*	African sheath-tailed bats	
C. afra	East African sheath-tailed bat	No assessment
C. seychellensis	Seychelles sheath-tailed bat	Endangered
Genus *Cormura*		
C. brevirostris	Wagner's sac-winged bat	Stable
Genus *Cyttarops*		
C. alecto	Short-eared bat	Stable
Genus *Diclidurus*	White bats	
D. albus	Northern white bat	Stable
D. ingens	Greater white bat	Stable
D. isabellus	Isabelle's white bat	Stable
D. scutatus	Lesser white bat	Stable
Genus *Emballonura*	Old World sheath-tailed bats	
E. alecto	Philippine sheath-tailed bat	No assessment
E. atrata	Peters's sheath-tailed bat	No assessment
E. beccarii	Beccari's sheath-tailed bat	No assessment
E. dianae	Rennell Island sheath-tailed bat	No assessment
E. furax	Greater sheath-tailed bat	Potentially vulnerable
E. monticola	Lesser sheath-tailed bat	No assessment
E. raffrayana	Raffray's sheath-tailed bat	Potentially vulnerable

Scientific Name	Common Name	Conservation Status
E. semicaudata	Polynesian sheath-tailed bat	Potentially vulnerable
Genus *Mosia*		
M. nigrescens	Dark sheath-tailed bat	No assessment
Genus *Peropteryx*	Doglike bats	
P. kappleri	Dusky doglike bat	Stable
P. leucoptera	White-winged doglike bat	Stable
P. macrotis	Lesser doglike bat	Stable
Genus *Rhynchonycteris*		
R. naso	Tufted bat	Potentially vulnerable
Genus *Saccolaimus*	Pouched bats	
S. flaviventris	Yellow-bellied pouched bat	No assessment
S. mixtus	Troughton's pouched bat	Potentially vulnerable
S. peli	Pel's pouched bat	No assessment
S. pluto	Philippine pouched bat	No assessment
S. saccolaimus	White-bellied pouch bat	Potentially vulnerable
Genus *Saccopteryx*	Sac-winged bats	
S. bilineata	Greater sac-winged bat	Stable
S. canescens	Frosted sac-winged bat	Stable
S. gymnura	Amazonian sac-winged bat	Stable
S. leptura	Lesser sac-winged bat	Stable
Genus *Taphozous*	Tomb bats	
T. australis	Gould's tomb bat	Potentially vulnerable
T. georgianus	Sharp-nosed tomb bat	No assessment
T. hamiltoni	Hamilton's tomb bat	No assessment
T. hildegardeae	Hildegarde's tomb bat	No assessment
T. hilli	Hill's tomb bat	No assessment
T. kapalgensis	White-striped tomb bat	Potentially vulnerable
T. longimanus	Long-winged tomb bat	No assessment
T. mauritianus	Mauritian tomb bat	No assessment
T. melanopogon	Black-bearded tomb bat	No assessment
T. nudiventris	Naked-rumped tomb bat	No assessment
T. perforatus	Egyptian tomb bat	No assessment
T. philippinensis	Philippine tomb bat	No assessment
T. theobaldi	Theobald's tomb bat	No assessment

Family Nycteridae

Genus *Nycteris*	Slit-faced bats	
N. arge	Bate's slit-faced bat	No assessment
N. gambiensis	Gambian slit-faced bat	No assessment
N. grandis	Large slit-faced bat	No assessment
N. hispida	Hairy slit-faced bat	No assessment
N. intermedia	Intermediate slit-faced bat	No assessment
N. javanica	Javan slit-faced bat	No assessment
N. macrotis	Large-eared slit-faced bat	No assessment
N. major	Ja slit-faced bat	No assessment
N. nana	Dwarf slit-faced bat	No assessment
N. thebaica	Egyptian slit-faced bat	No assessment
N. tragata	Malaysian slit-faced bat	No assessment
N. woodi	Wood's slit-faced bat	No assessment

Family Megadermatidae—false vampire and yellow-winged bats

Genus *Cardioderma*		
C. cor	Heart-nosed bat	No assessment
Genus *Lavia*		
L. frons	Yellow-winged bat	No assessment

Continued

Scientific Name	Common Name	Conservation Status

Family Megadermatidae—false vampire and yellow-winged bats *(continued)*

Scientific Name	Common Name	Conservation Status
Genus *Macroderma*		
M. *gigas*	Australian ghost bat	Vulnerable
Genus *Megaderma*	Asian false vampire bats	
M. *lyra*	Greater false vampire bat	No assessment
M. *spasma*	Malayan false vampire bat	No assessment

Family Rhinolophidae—horseshoe bats

Scientific Name	Common Name	Conservation Status
Genus *Rhinolophus*	Horseshoe bats	
R. *acuminatus*	Acuminate horseshoe bat	No assessment
R. *adami*	Adam's horseshoe bat	No assessment
R. *affinis*	Intermediate horseshoe bat	Stable
R. *alcyone*	Halcyon horseshoe bat	No assessment
R. *anderseni*	Andersen's horseshoe bat	No assessment
R. *arcuatus*	Arcuate horseshoe bat	No assessment
R. *blasii*	Blasius's horseshoe bat	Potentially vulnerable
R. *borneensis*	Bornean horseshoe bat	No assessment
R. *canuti*	Canut's horseshoe bat	No assessment
R. *capensis*	Cape horseshoe bat	No assessment
R. *celebensis*	Sulawesi horseshoe bat	No assessment
R. *clivosus*	Geoffroy's horseshoe bat	No assessment
R. *coelophyllus*	Croslet horseshoe bat	No assessment
R. *cognatus*	Kindred horseshoe bat	No assessment
R. *cornutus*	Little Japanese horseshoe bat	No assessment
R. *creaghi*	Creagh's horseshoe bat	No assessment
R. *darlingi*	Darling's horseshoe bat	No assessment
R. *deckenii*	Decken's horseshoe bat	No assessment
R. *denti*	Dent's horseshoe bat	No assessment
R. *eloquens*	Eloquent horseshoe bat	No assessment
R. *euryale*	Mediterranean horseshoe bat	Potentially vulnerable
R. *euryotis*	Broad-eared horseshoe bat	No assessment
R. *ferrumequinum*	Greater horseshoe bat	Vulnerable
R. *fumigatus*	Rüppell's horseshoe bat	No assessment
R. *guineensis*	Guinean horseshoe bat	No assessment
R. *hildebrandti*	Hildebrandt's horseshoe bat	No assessment
R. *hipposideros*	Lesser horseshoe bat	Vulnerable
R. *imaizumii*	Imaizumi's horseshoe bat	Potentially vulnerable
R. *inops*	Philippine forest horseshoe bat	No assessment
R. *keyensis*	Insular horseshoe bat	No assessment
R. *landeri*	Lander's horseshoe bat	No assessment
R. *lepidus*	Blyth's horseshoe bat	No assessment
R. *luctus*	Woolly horseshoe bat	No assessment
R. *maclaudi*	Maclaud's horseshoe bat	No assessment
R. *macrotis*	Big-eared horseshoe bat	No assessment
R. *malayanus*	Malayan horseshoe bat	No assessment
R. *marshalli*	Marshall's horseshoe bat	Potentially vulnerable
R. *megaphyllus*	Smaller horseshoe bat	No assessment
R. *mehelyi*	Mehely's horseshoe bat	Potentially vulnerable
R. *mitratus*	Indian horseshoe bat	No assessment
R. *monoceros*	Taiwan horseshoe bat	No assessment
R. *nereis*	Sea nymph horseshoe bat	No assessment
R. *osgoodi*	Osgood's horseshoe bat	No assessment
R. *paradoxolophus*	Bourret's horseshoe bat	Potentially vulnerable
R. *pearsoni*	Pearson's horseshoe bat	No assessment
R. *philippinensis*	Philippine horseshoe bat	Potentially vulnerable
R. *pusillus*	Least horseshoe bat	No assessment

Scientific Name	Common Name	Conservation Status
R. rex	King horseshoe bat	No assessment
R. robinsoni	Peninsular horseshoe bat	No assessment
R. rouxi	Rufous horseshoe bat	No assessment
R. rufus	Red horseshoe bat	No assessment
R. sedulus	Lesser woolly horseshoe bat	Potentially vulnerable
R. shameli	Shamel's horseshoe bat	No assessment
R. silvestris	Forest horseshoe bat	No assessment
R. simplex	Lombok horseshoe bat	No assessment
R. simulator	Bushveld horseshoe bat	No assessment
R. stheno	Lesser brown horseshoe bat	No assessment
R. subbadius	Chestnut horseshoe bat	No assessment
R. subrufus	Luzon horseshoe bat	No assessment
R. swinnyi	Swinny's horseshoe bat	No assessment
R. thomasi	Thomas's horseshoe bat	No assessment
R. trifoliatus	Trefoil horseshoe bat	No assessment
R. virgo	Yellow-faced horseshoe bat	No assessment
R. yunanensis	Dobson's horseshoe bat	No assessment
Genus Anthops		
A. ornatus	Flower-faced bat	Potentially vulnerable
Genus Asellia	Trident leaf-nosed bats	
A. patrizii	Patrizi's trident leaf-nosed bat	No assessment
A. tridens	Trident leaf-nosed bat	No assessment
Genus Aselliscus	Tate's trident-nosed bats	
A. stoliczkanus	Stoliczka's trident bat	No assessment
A. tricuspidatus	Temminck's trident bat	No assessment
Genus Cloeotis		
C. percivali	Percival's trident bat	Potentially vulnerable
Genus Coelops	Tailless leaf-nosed bats	
C. frithi	East Asian tailless leaf-nosed bat	Potentially vulnerable
C. hirsutus	Hairy tailless leaf-nosed bat	No assessment
C. robinsoni	Malayan tailless leaf-nosed bat	No assessment
Genus Hipposideros	Roundleaf bats	
H. abae	Aba roundleaf bat	No assessment
H. armiger	Himalayan roundleaf bat	No assessment
H. ater	Dusky roundleaf bat	No assessment
H. beatus	Benito roundleaf bat	No assessment
H. bicolor	Bicolored roundleaf bat	No assessment
H. breviceps	Short-headed roundleaf bat	No assessment
H. caffer	Sundevall's roundleaf bat	No assessment
H. calcaratus	Spurred roundleaf bat	No assessment
H. camerunensis	Greater roundleaf bat	No assessment
H. cervinus	Fawn roundleaf bat	No assessment
H. cineraceus	Ashy roundleaf bat	No assessment
H. commersoni	Commerson's roundleaf bat	Stable
H. coronatus	Large Mindanao roundleaf bat	No assessment
H. corynophyllus	Telefomin roundleaf bat	No assessment
H. coxi	Cox's roundleaf bat	Potentially vulnerable
H. crumeniferus	Timor roundleaf bat	Potentially vulnerable
H. curtus	Short-tailed roundleaf bat	No assessment
H. cyclops	Cyclops roundleaf bat	No assessment
H. diadema	Diadem roundleaf bat	No assessment
H. dinops	Fierce roundleaf bat	No assessment
H. doriae	Borneo roundleaf bat	No assessment
H. dyacorum	Dayak roundleaf bat	No assessment
H. fuliginosus	Sooty roundleaf bat	No assessment
H. fulvus	Fulvous roundleaf bat	No assessment
H. galeritus	Cantor's roundleaf bat	Potentially vulnerable

Continued

Scientific Name	Common Name	Conservation Status
Family Rhinolophidae—horseshoe bats (*continued*)		
H. halophyllus	Thailand roundleaf bat	No assessment
H. inexpectatus	Crested roundleaf bat	No assessment
H. jonesi	Jones's roundleaf bat	No assessment
H. lamottei	Lamotte's roundleaf bat	No assessment
H. lankadiva	Indian roundleaf bat	No assessment
H. larvatus	Horsfield's roundleaf bat	No assessment
H. lekaguli	Large Asian roundleaf bat	Potentially vulnerable
H. lylei	Shield-faced roundleaf bat	No assessment
H. macrobullatus	Big-eared roundleaf bat	No assessment
H. maggietaylorae	Maggie's roundleaf bat	No assessment
H. marisae	Aellen's roundleaf bat	No assessment
H. megalotis	Ethiopian large-eared roundleaf bat	No assessment
H. muscinus	Fly-river roundleaf bat	No assessment
H. nequam	Malayan roundleaf bat	No assessment
H. obscurus	Philippine forest roundleaf bat	No assessment
H. papua	Geelvink Bay roundleaf bat	Potentially vulnerable
H. pomona	Pomona roundleaf bat	No assessment
H. pratti	Pratt's roundleaf bat	No assessment
H. pygmaeus	Philippine pygmy roundleaf bat	No assessment
H. ridleyi	Ridley's roundleaf bat	Potentially vulnerable
H. ruber	Noack's roundleaf bat	Stable
H. sabanus	Least roundleaf bat	No assessment
H. schistaceus	Split roundleaf bat	No assessment
H. semoni	Semon's roundleaf bat	Potentially vulnerable
H. speoris	Schneider's roundleaf bat	No assessment
H. stenotis	Narrow-eared roundleaf bat	Potentially vulnerable
H. turpis	Lesser roundleaf bat	Potentially vulnerable
H. wollastoni	Wollaston's roundleaf bat	Potentially vulnerable
Genus *Paracoelops*		
P. megalotis	Vietnam leaf-nosed bat	Potentially vulnerable
Genus *Rhinonicteris*		
R. aurantia	Orange leaf-nosed bat	Potentially vulnerable
Genus *Triaenops*	Triple nose-leaf bats	
T. furculus	Trouessart's trident bat	No assessment
T. persicus	Persian trident bat	Vulnerable
Family Noctilionidae		
Genus *Noctilio*	Bulldog bats	
N. albiventris	Southern bulldog bat	Stable
N. leporinus	Greater bulldog bat	Potentially vulnerable
Family Mormoopidae—naked-backed and mustached bats		
Genus *Mormoops*	Leaf-chinned bats	
M. blainvillii	Blainville's leaf-chinned bat	Potentially vulnerable
M. megalophylla	Peters's leaf-chinned bat	Potentially vulnerable
Genus *Pteronotus*	Naked-backed and mustached bats	
P. davyi	Davy's naked-backed bat	Stable
P. gymnonotus	Big naked-backed bat	Stable
P. macleayii	Mac Leay's mustached bat	Potentially vulnerable
P. parnellii	Parnell's mustached bat	Stable
P. personatu	Wagner's mustached bat	Stable
P. quadridens	Sooty mustached bat	Stable

Scientific Name	Common Name	Conservation Status
Family Phyllostomidae—New World leaf-nosed bats		
Genus *Chrotopterus*		
C. *auritus*	Giant woolly bat	Potentially vulnerable
Genus *Lonchorhina*	Sword-nosed bats	
L. *aurita*	Tomes's sword-nosed bat	Potentially vulnerable
L. *fernandezi*	Fernandez's sword-nosed bat	Potentially vulnerable
L. *marinkellei*	Marinkelle's sword-nosed bat	Potentially vulnerable
L. *orinocensis*	Orinoco sword-nosed bat	Potentially vulnerable
Genus *Macrophyllum*		
M. *macrophyllum*	Long-legged bat	Stable
Genus *Macrotus*	Leaf-nosed bats	
M. *californicus*	California leaf-nosed bat	Potentially vulnerable
M. *waterhousii*	Waterhouse's leaf-nosed bat	Stable
Genus *Micronycteris*	Little big-eared bats	
M. *behnii*	Behni's big-eared bat	Potentially vulnerable
M. *brachyotis*	Yellow-throated big-eared bat	Stable
M. *daviesi*	Davies's big-eared bat	Potentially vulnerable
M. *hirsuta*	Hairy big-eared bat	Stable
M. *megalotis*	Little big-eared bat	Stable
M. *minuta*	White-bellied big-eared bat	Potentially vulnerable
M. *nicefori*	Nicefero's big-eared bat	Stable
M. *pusilla*	Least big-eared bat	Potentially vulnerable
M. *schmidtorum*	Schmidt's big-eared bat	Stable
M. *sylvestris*	Tricolored big-eared bat	Potentially vulnerable
Genus *Mimon*	Hairy-nosed bats	
M. *bennettii*	Bennett's hairy-nosed bat	Potentially vulnerable
M. *crenulatum*	Striped hairy-nosed bat	Stable
Genus *Phylloderma*		
P. *stenops*	Pale-faced spear-nosed bat	Potentially vulnerable
Genus *Phyllostomus*	Spear-nosed bats	
P. *discolor*	Pale spear-nosed bat	Stable
P. *elongatus*	Lesser spear-nosed bat	Stable
P. *hastatus*	Pallas's spear-nosed bat	Stable
P. *latifolius*	Guianan spear-nosed bat	Potentially vulnerable
Genus *Tonatia*	Round-eared bats	
T. *bidens*	Greater round-eared bat	Stable
T. *brasiliense*	Pygmy round-eared bat	Stable
T. *carrikeri*	Carriker's round-eared bat	Potentially vulnerable
T. *evotis*	Davis's round-eared bat	Potentially vulnerable
T. *schulzi*	Schultz's round-eared bat	Potentially vulnerable
T. *silvicola*	D'Orbigny's round-eared bat	Stable
Genus *Trachops*		
T. *cirrhosus*	Fringe-lipped bat	Stable
Genus *Vampyrum*		
V. *spectrum*	American false vampire bat	Potentially vulnerable
Genus *Lionycteris*		
L. *spurrelli*	Chestnut long-tongued bat	Potentially vulnerable
Genus *Lonchophylla*	Long-tongued bats	
L. *bokermanni*	Bokermann's long-tongued bat	Potentially vulnerable
L. *dekeyseri*	Dekeyser's long-tongued bat	Potentially vulnerable
L. *handleyi*	Handley's long-tongued bat	Potentially vulnerable
L. *hesperia*	Western long-tongued bat	Potentially vulnerable
L. *mordax*	Brazilian long-tongued bat	Potentially vulnerable
L. *robusta*	Panama long-tongued bat	Potentially vulnerable
L. *thomasi*	Thomas's long-tongued bat	Potentially vulnerable

Continued

Scientific Name	Common Name	Conservation Status
Family Phyllostomidae—New World leaf-nosed bats (*continued*)		
Genus *Platalina*		
P. genovensium	Long-snouted bat	Potentially vulnerable
Genus *Brachyphylla*	West Indian fruit-eating bats	
B. cavernarum	Antillean fruit-eating bat	Stable
B. nana	Cuban fruit-eating bat	Potentially vulnerable
Genus *Erophylla*		
E. sezekorni	Buffy flower bat	Potentially vulnerable
Genus *Phyllonycteris*	Smooth-toothed flower bats	
P. aphylla	Jamaican flower bat	Potentially vulnerable
P. poeyi	Cuban flower bat	Stable
Genus *Anoura*	Tailless bats	
A. caudifer	Tailed tailless bat	Stable
A. cultrata	Handley's tailless bat	Stable
A. geoffroyi	Geoffroy's tailless bat	Stable
A. latidens	Broad-toothed tailless bat	Potentially vulnerable
Genus *Choeroniscus*	Long-tailed bats	
C. godmani	Godman's long-tailed bat	Stable
C. intermedius	Intermediate long-tailed bat	Stable
C. minor	Least long-tailed bat	Stable
C. periosus	Immense long-tailed bat	Stable
Genus *Choeronycteris*		
C. mexicana	Mexican long-tongued bat	Potentially vulnerable
Genus *Glossophaga*	Nectar bats	
G. commissarisi	Commissaris's nectar bat	Stable
G. leachii	Leach's nectar bat	Stable
G. longirostris	Greater nectar bat	Stable
G. morenoi	Moreno's nectar bat	Stable
G. soricina	Long-tongued nectar bat	Stable
Genus *Hylonycteris*		
H. underwoodi	Underwood's long-tongued bat	Stable
Genus *Leptonycteris*	Long-nosed bats	
L. curasoae	Southern long-nosed bat	Vulnerable
L. nivalis	Mexican long-nosed bat	Vulnerable
Genus *Lichonycteris*		
L. obscura	Brown long-nosed bat	Potentially vulnerable
Genus *Monophyllus*	Single-leaf bats	
M. plethodon	Insular single leaf bat	Stable
M. redmani	Leach's single leaf bat	Potentially vulnerable
Genus *Musonycteris*		
M. harrisoni	Trumpet-nosed bat	Potentially vulnerable
Genus *Scleronycteris*		
S. ega	Ega long-tongued bat	Potentially vulnerable
Genus *Carollia*	Short-tailed bats	
C. brevicauda	Silky short-tailed bat	Stable
C. castanea	Chestnut short-tailed bat	Stable
C. perspicillata	Linnaeus's short-tailed bat	Stable
C. subrufa	Reddish short-tailed bat	Stable
Genus *Rhinophylla*	Little fruit bats	
R. alethina	Genuine little fruit bat	Potentially vulnerable
R. fischerae	Fischer's little fruit bat	Potentially vulnerable
R. pumilio	Dwarf little fruit bat	Stable
Genus *Ametrida*		
A. centurio	White-shouldered bat	Stable
Genus *Ardops*		
A. nichollsi	Tree bat	Stable

Scientific Name	Common Name	Conservation Status
Genus *Ariteus*		
A. flavescens	Jamaican fig-eating bat	Stable
Genus *Artibeus*	Neotropical fruit bats	
A. amplus	Large fruit bat	Stable
A. anderseni	Andersen's fruit bat	Stable
A. aztecus	Highland fruit bat	Stable
A. cinereus	Gervais's fruit bat	Stable
A. concolor	Brown fruit bat	Stable
A. fimbriatus	Fringed fruit bat	Stable
A. fraterculus	Fraternal fruit bat	Stable
A. glaucus	Silver fruit bat	Stable
A. hartii	Hart's fruit bat	Stable
A. hirsutus	Hairy fruit bat	Stable
A. inopinatus	Honduran fruit bat	Stable
A. jamaicensis	Jamaican fruit bat	Stable
A. lituratus	Big fruit bat	Stable
A. obscurus	Dark fruit bat	Stable
A. phaeotis	Dwarf fruit bat	Stable
A. planirostris	Flat-faced fruit bat	Stable
A. toltecus	Lowland fruit bat	Stable
Genus *Centurio*		
C. senex	Wrinkle-faced bat	Stable
Genus *Chiroderma*	Big-eyed bats	
C. doriae	Brazilian big-eyed bat	Stable
C. improvisum	Guadeloupe big-eyed bat	Potentially vulnerable
C. salvini	Salvin's big-eyed bat	Stable
C. trinitatum	Goodwin's big-eyed bat	Stable
C. villosum	Greater big-eyed bat	Stable
Genus *Ectophylla*		
E. alba	White fruit bat	Stable
Genus *Mesophylla*		
M. macconnelli	Macconnell's bat	Stable
Genus *Phyllops*		
P. falcatus	Cuban fig-eating bat	Stable
Genus *Platyrrhinus*	Broad-nosed bats	
P. aurarius	Eldorado broad-nosed bat	Stable
P. brachycephalus	Short-headed broad-nosed bat	Stable
P. chocoensis	Choco broad-nosed bat	Stable
P. dorsalis	Thomas's broad-nosed bat	Stable
P. helleri	Heller's broad-nosed bat	Stable
P. infuscus	Buffy broad-nosed bat	Stable
P. lineatus	White-lined broad-nosed bat	Stable
P. recifinus	Recife broad-nosed bat	Stable
P. umbratus	Shadowy broad-nosed bat	Stable
P. vittatus	Greater broad-nosed bat	Stable
Genus *Pygoderma*		
P. bilabiatum	Ipanema bat	Potentially vulnerable
Genus *Sphaeronycteris*		
S. toxophyllum	Visored bat	Potentially vulnerable
Genus *Stenoderma*		
S. rufum	Red fruit bat	Potentially vulnerable
Genus *Sturnira*	American epauletted bats	
S. aratathomasi	Andes epauletted bat	Stable
S. bidens	Bidentate epauletted bat	Stable
S. bogotensis	Bogota epauletted bat	Stable
S. erythromos	Reddish epauletted bat	Stable
S. lilium	Yellow epauletted bat	Stable

Continued

Scientific Name	Common Name	Conservation Status
Family Phyllostomidae—New World leaf-nosed bats (*continued*)		
S. ludovici	Ludovic's epauletted bat	Stable
S. luisi	Luis's epauletted bat	Stable
S. magna	Greater epauletted bat	Stable
S. mordax	Hairy-footed epauletted bat	Stable
S. nana	Lesser epauletted bat	Stable
S. thomasi	Thomas's epauletted bat	Stable
S. tildae	Tilda epauletted bat	Stable
Genus *Uroderma*	Tent-making bats	
U. bilobatum	Tent-making bat	Stable
U. magnirostrum	Big-nosed tent-making bat	Stable
Genus *Vampyressa*	Yellow-eared bats	
V. bidens	Bidentate yellow-eared bat	Stable
V. brocki	Brock's yellow-eared bat	Stable
V. melissa	Melissa's yellow-eared bat	Stable
V. nymphaea	Big yellow-eared bat	Stable
V. pusilla	Little yellow-eared bat	Stable
Genus *Vampyrodes*		
V. caraccioli	Great stripe-faced bat	Stable
Genus *Desmodus*		
D. rotundus	Common vampire bat	Stable
Genus *Diaemus*		
D. youngi	White-winged vampire bat	Stable
Genus *Diphylla*		
D. ecaudata	Hairy-legged vampire bat	Stable
Family Natalidae		
Genus *Natalus*	Funnel-eared bats	
N. lepidus	Gervais's funnel-eared bat	Potentially vulnerable
N. micropus	Cuban funnel-eared bat	Potentially vulnerable
N. stramineus	Mexican funnel-eared bat	Stable
N. tumidifrons	Bahaman funnel-eared bat	Potentially vulnerable
N. tumidirostris	Trinidadian funnel-eared bat	Stable
Family Furipteridae—smoky and thumbless bats		
Genus *Amorphochilus*		
A. schnablii	Smoky bat	Potentially vulnerable
Genus *Furipterus*		
F. horrens	Thumbless bat	Potentially vulnerable
Family Thyropteridae		
Genus *Thyroptera*	Disk-winged bats	
T. discifera	Peters's disk-winged bat	Stable
T. tricolor	Spix's disk-winged bat	Stable
Family Myzopodidae		
Genus *Myzopoda*		
M. aurita	Sucker-footed bat	Vulnerable
Family Vespertilionidae—vespertilionid bats		
Genus *Kerivoula*	Woolly bats	
K. aerosa	Dubious trumpet-eared bat	No assessment
K. africana	Tanzanian woolly bat	Probably extinct
K. agnella	St. Aignan's trumpet-eared bat	No assessment
K. argentata	Damara woolly bat	No assessment

Scientific Name	Common Name	Conservation Status
K. atrox	Groove-toothed bat	Potentially vulnerable
K. cuprosa	Copper woolly bat	No assessment
K. eriophora	Ethiopian woolly bat	No assessment
K. flora	Flores woolly bat	No assessment
K. hardwickei	Hardwick's woolly bat	Potentially vulnerable
K. intermedia	Small woolly bat	No assessment
K. jagori	Peters's trumpet-eared bat	Potentially vulnerable
K. lanosa	Lesser woolly bat	No assessment
K. minuta	Least woolly bat	No assessment
K. muscina	Fly River trumpet-eared bat	No assessment
K. myrella	Bismarck's trumpet-eared bat	No assessment
K. papillosa	Papillose woolly bat	Potentially vulnerable
K. papuensis	Papuan trumpet-eared bat	No assessment
K. pellucida	Clear-winged woolly bat	No assessment
K. phalaena	Spurrell's woolly bat	No assessment
K. picta	Painted bat	Potentially vulnerable
K. smithi	Smith's woolly bat	No assessment
K. whiteheadi	Whitehead's woolly bat	No assessment
Genus *Antrozous*	Pallid bats	
A. dubiaquercus	Van Gelder's bat	Stable
A. pallidus	Pallid bat	Stable
Genus *Barbastella*	Barbastelles	
B. barbastellus	Western barbastelle	No assessment
B. leucomelas	Eastern barbastelle	No assessment
Genus *Chalinolobus*	Wattled bats	
C. alboguttatus	Allen's striped bat	No assessment
C. argentatus	Silvered bat	No assessment
C. beatrix	Beatrix's bat	No assessment
C. dwyeri	Large-eared wattled bat	No assessment
C. egeria	Bibundi wattled bat	Potentially vulnerable
C. gleni	Glen's wattled bat	No assessment
C. gouldii	Gould's wattled bat	No assessment
C. kenyacola	Kenya wattled bat	No assessment
C. morio	Chocolate wattled bat	No assessment
C. nigrogriseus	Frosted wattled bat	No assessment
C. picatus	Little pied wattled bat	No assessment
C. poensis	Abo bat	No assessment
C. superbus	Pied wattled bat	Potentially vulnerable
C. tuberculatus	Long-tailed wattled bat	Potentially vulnerable
C. variegatus	Butterfly bat	No assessment
Genus *Eptesicus*	Serotines	
E. baverstocki	Baverstock's serotine	No assessment
E. bobrinskoi	Bobrinskoy's serotine	No assessment
E. bottae	Botta's serotine	No assessment
E. brasiliensis	Brazilian brown bat	Stable
E. brunneus	Dark-brown serotine	No assessment
E. capensis	Cape serotine	No assessment
E. demissus	Surat serotine	Potentially vulnerable
E. diminutus	Diminutive serotine	Stable
E. douglasorum	Yellow-lipped bat	Potentially vulnerable
E. flavescens	Yellow serotine	No assessment
E. floweri	Horn-skinned bat	No assessment
E. furinalis	Tropical brown bat	Stable
E. fuscus	Big brown bat	Stable
E. guadeloupensis	Guadeloupe brown bat	Potentially vulnerable
E. guineensis	Tiny serotine	No assessment
E. hottentotus	Long-tailed house bat	No assessment

Continued

Scientific Name	Common Name	Conservation Status
Family Vespertilionidae—vespertilionid bats (*continued*)		
E. innoxius	Harmless serotine	Stable
E. kobayashii	Kobayashi's serotine	No assessment
E. melckorum	Melck's house bat	No assessment
E. nasutus	Sind bat	Potentially vulnerable
E. nilssoni	Northern bat	Potentially vulnerable
E. pachyotis	Thick-eared bat	No assessment
E. platyops	Lagos serotine	Potentially vulnerable
E. pumilus	Little bat	No assessment
E. regulus	King River bat	No assessment
E. rendalli	Rendall's serotine	No assessment
E. sagittula	Large forest bat	No assessment
E. serotinus	Serotine	Potentially vulnerable
E. somalicus	Somali serotine	No assessment
E. tatei	Tate's serotine	No assessment
E. tenuipinnis	White-winged serotine	No assessment
E. vulturnus	Little forest bat	No assessment
Genus *Euderma*		
E. maculatum	Spotted bat	Stable
Genus *Eudiscopus*		
E. denticulus	Disk-footed bat	Potentially vulnerable
Genus *Glischropus*	Thick-thumbed bats	
G. javanus	Javan thick-thumbed bat	Potentially vulnerable
G. tylopus	Common thick-thumbed bat	No assessment
Genus *Hesperoptenus*	False serotines	
H. blanfordi	Blanford's bat	No assessment
H. doriae	False serotine bat	Potentially vulnerable
H. gaskelli	Gaskell's false serotine	No assessment
H. tickelli	Tickell's bat	No assessment
H. tomesi	Large false serotine	No assessment
Genus *Histiotus*	Big-eared brown bats	
H. alienus	Strange big-eared brown bat	Stable
H. macrotus	Big-eared brown bat	Stable
H. montanus	Small big-eared brown bat	Stable
H. velatus	Tropical big-eared brown bat	Stable
Genus *Ia*		
I. io	Great evening bat	No assessment
Genus *Idionycteris*		
I. phyllotis	Allen's big-eared bat	Stable
Genus *Laephotis*	African long-eared bats	
L. angolensis	Angola long-eared bat	Potentially vulnerable
L. botswanae	Botswana long-eared bat	Potentially vulnerable
L. namibensis	Namib long-eared bat	Potentially vulnerable
L. wintoni	De Winton's long-eared bat	Potentially vulnerable
Genus *Lasionycteris*		
L. noctivagans	Silver-haired bat	Stable
Genus *Lasiurus*	Hairy-tailed bats	
L. borealis	Red bat	Stable
L. castaneus	Tacarcuna bat	Stable
L. cinereus	Hoary bat	Stable
L. ega	Southern yellow bat	Stable
L. egregius	Big red bat	Stable
L. intermedius	Northern yellow bat	Stable
L. seminolus	Seminole bat	Stable
Genus *Mimetillus*		
M. moloneyi	Moloney's flat-headed bat	No assessment

Scientific Name	Common Name	Conservation Status
Genus Myotis	Little brown bats	
M. abei	Sakhalin myotis	Potentially vulnerable
M. adversus	Large-footed bat	No assessment
M. aelleni	Southern myotis	Stable
M. albescens	Silver-tipped myotis	Stable
M. altarium	Szechwan myotis	No assessment
M. annectans	Hairy-faced bat	No assessment
M. atacamensis	Atacama myotis	Stable
M. auriculus	Southwestern myotis	Stable
M. australis	Australian myotis	No assessment
M. austroriparius	Southeastern myotis	Vulnerable
M. bechsteini	Bechstein's bat	Vulnerable
M. blythii	Lesser mouse-eared bat	Vulnerable
M. bocagei	Rufous mouse-eared bat	No assessment
M. bombinus	Far Eastern myotis	No assessment
M. brandti	Brandt's bat	No assessment
M. californicus	California myotis	Stable
M. capaccinii	Long-fingered bat	Vulnerable
M. chiloensis	Chilean myotis	Stable
M. chinensis	Large myotis	No assessment
M. cobanensis	Guatemalan myotis	Endangered
M. dasycneme	Pond bat	Potentially vulnerable
M. daubentoni	Daubenton's bat	Potentially vulnerable
M. dominicensis	Dominican myotis	Stable
M. elegans	Elegant myotis	Stable
M. emarginatus	Geoffroy's bat	No assessment
M. evotis	Long-eared myotis	Stable
M. findleyi	Findley's myotis	Stable
M. formosus	Hodgson's bat	Stable
M. fortidens	Cinnamon myotis	Stable
M. frater	Fraternal myotis	No assessment
M. goudoti	Malagasy mouse-eared bat	No assessment
M. grisescens	Gray bat	Endangered
M. hasseltii	Lesser large-footed bat	Potentially vulnerable
M. horsfieldii	Horsfield's bat	No assessment
M. hosonoi	Hosono's myotis	No assessment
M. ikonnikovi	Ikonnikow's bat	No assessment
M. insularum	Insular myotis	Stable
M. keaysi	Hairy-legged myotis	Stable
M. keenii	Keen's myotis	Stable
M. leibii	Eastern small-footed myotis	Stable
M. lesueuri	Lesueur's hairy bat	Potentially vulnerable
M. levis	Yellowish myotis	Stable
M. longipes	Long-footed myotis	No assessment
M. lucifugus	Little brown bat	Stable
M. macrodactylus	Big-footed myotis	No assessment
M. macrotarsus	Pallid large-footed myotis	No assessment
M. martiniquensis	Schwartz's myotis	Stable
M. milleri	Miller's myotis	Endangered
M. montivagus	Burmese whiskered bat	No assessment
M. morrisi	Morris's bat	No assessment
M. muricola	Whiskered myotis	No assessment
M. myotis	Mouse-eared bat	Potentially vulnerable
M. mystacinus	Whiskered bat	No assessment
M. nattereri	Natterer's bat	No assessment
M. nesopolus	Curacao myotis	Stable
M. nigricans	Black myotis	Stable

Continued

Scientific Name	Common Name	Conservation Status
Family Vespertilionidae—vespertilionid bats *(continued)*		
M. *oreias*	Singapore whiskered bat	No assessment
M. *oxyotus*	Montane myotis	Stable
M. *ozensis*	Honshu myotis	No assessment
M. *peninsularis*	Peninsular myotis	Stable
M. *pequinius*	Peking myotis	No assessment
M. *planiceps*	Flat-headed myotis	Endangered
M. *pruinosus*	Frosted myotis	No assessment
M. *ricketti*	Rickett's big-footed bat	No assessment
M. *ridleyi*	Ridley's bat	No assessment
M. *riparius*	Riparian myotis	Stable
M. *rosseti*	Thick-thumbed myotis	No assessment
M. *ruber*	Red myotis	Stable
M. *schaubi*	Schaub's myotis	No assessment
M. *scotti*	Scott's mouse-eared bat	No assessment
M. *seabrai*	Angola hairy bat	Potentially vulnerable
M. *sicarius*	Assassin bat	No assessment
M. *siligorensis*	Himalayan whiskered bat	No assessment
M. *simus*	Velvety myotis	Stable
M. *sodalis*	Indiana bat	Endangered
M. *stalkeri*	Kei myotis	No assessment
M. *thysanodes*	Fringed myotis	Stable
M. *tricolor*	Cape hairy bat	Potentially vulnerable
M. *velifer*	Cave myotis	Endangered
M. *vivesi*	Fish-eating bat	Stable
M. *volans*	Long-legged myotis	Stable
M. *welwitschii*	Welwitch's hairy bat	No assessment
M. *yesoensis*	Yoshiyuki's myotis	No assessment
M. *yumanensis*	Yuma myotis	Stable
Genus *Nyctalus*	Noctule bats	
N. *aviator*	Birdlike noctule	No assessment
N. *azoreum*	Azores noctule	No assessment
N. *lasiopterus*	Giant noctule	Vulnerable
N. *leisleri*	Hairy-sided noctule	Potentially vulnerable
N. *montanus*	Montane noctule	No assessment
N. *noctula*	Noctule	No assessment
Genus *Nycticeius*	Evening bats	
N. *balstoni*	Western broad-nosed bat	No assessment
N. *greyii*	Little broad-nosed bat	No assessment
N. *humeralis*	Evening bat	Stable
N. *rueppellii*	Rüppell's broad-nosed bat	No assessment
N. *sanborni*	Northern broad-nosed bat	No assessment
N. *schlieffeni*	Schlieffen's bat	No assessment
Genus *Nyctophilus*	Long-eared bats	
N. *arnhemensis*	Arnhem long-eared bat	No assessment
N. *geoffroyi*	Geoffroy's long-eared bat	No assessment
N. *heran*	Sunda long-eared bat	No assessment
N. *microdon*	Small-toothed long-eared bat	No assessment
N. *microtis*	Papuan long-eared bat	Potentially vulnerable
N. *timoriensis*	Australian long-eared bat	No assessment
N. *walkeri*	Pygmy long-eared bat	Potentially vulnerable
Genus *Otonycteris*		
O. *hemprichi*	Hemprich's long-eared bat	No assessment
Genus *Pharotis*		
P. *imogene*	Imogene bat	Potentially vulnerable
Genus *Philetor*		
P. *brachypterus*	Rohu's bat	Potentially vulnerable

Scientific Name	Common Name	Conservation Status
Genus *Pipistrellus*	Pipistrelles	
P. aegyptius	Egyptian pipistrelle	No assessment
P. aero	Mt. Gargues pipistrelle	Potentially vulnerable
P. affinis	Chocolate bat	No assessment
P. anchietai	Anchieta's pipistrelle	No assessment
P. anthonyi	Anthony's pipistrelle	Potentially vulnerable
P. arabicus	Arabian pipistrelle	Potentially vulnerable
P. ariel	Desert pipistrelle	No assessment
P. babu	Himalayan pipistrelle	No assessment
P. bodenheimeri	Bodenheimer's pipistrelle	Potentially vulnerable
P. cadornae	Cadorna's pipistrelle	No assessment
P. ceylonicus	Kelaart's pipistrelle	No assessment
P. circumdatus	Black gilded pipistrelle	Potentially vulnerable
P. coromandra	Indian pipistrelle	No assessment
P. crassulus	Broad-headed pipistrelle	No assessment
P. cuprosus	Coppery pipistrelle	Potentially vulnerable
P. dormeri	Dormer's pipistrelle	No assessment
P. eisentrauti	Eisentraut's pipistrelle	No assessment
P. endoi	Endo's pipistrelle	No assessment
P. hesperus	Western pipistrelle	Stable
P. imbricatus	Brown pipistrelle	No assessment
P. inexspectatus	Aellen's pipistrelle	No assessment
P. javanicus	Javan pipistrelle	Stable
P. joffrei	Joffre's pipistrelle	Potentially vulnerable
P. kitcheneri	Red-brown pipistrelle	Potentially vulnerable
P. kuhlii	Kuhl's pipistrelle	Potentially vulnerable
P. lophurus	Burma pipistrelle	No assessment
P. macrotis	Big-eared pipistrelle	Potentially vulnerable
P. maderensis	Madeira pipistrelle	Potentially vulnerable
P. mimus	Indian pygmy pipistrelle	No assessment
P. minahassae	Minahassa pipistrelle	Potentially vulnerable
P. mordax	Pungent pipistrelle	Potentially vulnerable
P. musciculus	Mouselike pipistrelle	No assessment
P. nanulus	Tiny pipistrelle	No assessment
P. nanus	Banana bat	No assessment
P. nathusii	Nathusius's pipistrelle	Potentially vulnerable
P. paterculus	Paternal pipistrelle	No assessment
P. peguensis	Pegu pipistrelle	No assessment
P. permixtus	Dar-es-Salaam pipistrelle	Potentially vulnerable
P. petersi	Peters's pipistrelle	No assessment
P. pipistrellus	Common pipistrelle	Potentially vulnerable
P. pulveratus	Chinese pipistrelle	No assessment
P. rueppelli	Rüppell's pipistrelle	No assessment
P. rusticus	Rusty pipistrelle	No assessment
P. savii	Savi's pipistrelle	Potentially vulnerable
P. societatis	Social pipistrelle	Potentially vulnerable
P. stenopterus	Narrow-winged pipistrelle	No assessment
P. sturdeei	Sturdee's pipistrelle	No assessment
P. subflavus	Eastern pipistrelle	Stable
P. tasmaniensis	False pipistrelle	No assessment
P. tenuis	Least pipistrelle	Potentially vulnerable
Genus *Plecotus*	Big-eared bats	
P. auritus	Brown big-eared bat	Potentially vulnerable
P. austriacus	Gray big-eared bat	Potentially vulnerable
P. mexicanus	Mexican big-eared bat	Stable
P. rafinesquii	Rafinesque's big-eared bat	Endangered
P. taivanus	Taiwan big-eared bat	No assessment

Continued

Scientific Name	Common Name	Conservation Status
Family Vespertilionidae—vespertilionid bats (*continued*)		
P. teneriffae	Canary big-eared bat	No assessment
P. townsendii	Townsend's big-eared bat	Vulnerable
Genus *Rhogeessa*	Little yellow bats	
R. alleni	Allen's yellow bat	Stable
R. genowaysi	Genoways's yellow bat	Potentially vulnerable
R. gracilis	Slender yellow bat	Stable
R. minutilla	Tiny yellow bat	Stable
R. mira	Least yellow bat	Potentially vulnerable
R. parvula	Little yellow bat	Stable
R. tumida	Central American yellow bat	Stable
Genus *Scotoecus*	House bats	
S. albofuscus	Gambian light-winged house bat	No assessment
S. hirundo	Swallowlike house bat	No assessment
S. pallidus	Pallid house bat	No assessment
Genus *Scotomanes*	Harlequin bats	
S. emarginatus	Emarginate harlequin bat	Potentially vulnerable
S. ornatus	Harlequin bat	No assessment
Genus *Scotophilus*	Yellow bats	
S. borbonicus	Lesser yellow house bat	Probably extinct
S. celebensis	Sulawesi yellow bat	No assessment
S. dinganii	African yellow bat	No assessment
S. heathi	Asiatic greater yellow bat	No assessment
S. kuhlii	Asiatic lesser yellow bat	No assessment
S. leucogaster	White-bellied yellow bat	No assessment
S. nigrita	Schreber's yellow bat	No assessment
S. nux	Nut-colored yellow bat	No assessment
S. robustus	Robust yellow bat	No assessment
S. viridis	Greenish yellow bat	No assessment
Genus *Tylonycteris*	Bamboo bats	
T. pachypus	Lesser bamboo bat	No assessment
T. robustula	Greater bamboo bat	No assessment
Genus *Vespertilio*	Particolored bats	
V. murinus	Particolored bat	No assessment
V. superans	Asian particolored bat	No assessment
Genus *Harpiocephalus*		
H. harpia	Hairy-winged bat	Potentially vulnerable
Genus *Murina*	Tube-nosed insectivorous bats	
M. aenea	Bronze tube-nosed bat	No assessment
M. aurata	Little tube-nosed bat	No assessment
M. cyclotis	Round-eared tube-nosed bat	No assessment
M. florium	Flores tube-nosed bat	Potentially vulnerable
M. fusca	Dusky tube-nosed bat	No assessment
M. grisea	Peters's tube-nosed bat	No assessment
M. huttoni	Hutton's tube-nosed bat	No assessment
M. leucogaster	Greater tube-nosed bat	No assessment
M. puta	Taiwan tube-nosed bat	Potentially vulnerable
M. rozendaali	Gilded tube-nosed bat	No assessment
M. silvatica	Forest tube-nosed bat	Potentially vulnerable
M. suilla	Brown tube-nosed bat	Potentially vulnerable
M. tenebrosa	Gloomy tube-nosed bat	Potentially vulnerable
M. tubinaris	Pakistan tube-nosed bat	No assessment
M. ussuriensis	Ussuri tube-nosed bat	No assessment
Genus *Miniopterus*	Long-fingered bats	
M. australis	Little long-fingered bat	Vulnerable
M. fraterculus	Lesser long-fingered bat	No assessment

Scientific Name	Common Name	Conservation Status
M. *fuscus*	Southeast Asian long-fingered bat	No assessment
M. *inflatus*	Greater long-fingered bat	No assessment
M. *magnater*	Western bentwing bat	No assessment
M. *minor*	Least long-fingered bat	No assessment
M. *pusillus*	Small bent-winged bat	No assessment
M. *robustior*	Loyalty bent-winged bat	Potentially vulnerable
M. *schreibersi*	Schreibers's long-fingered bat	Potentially vulnerable
M. *tristis*	Great bentwing bat	Potentially vulnerable
Genus *Tomopeas*		
T. *ravus*	Blunt-eared bat	Potentially vulnerable

Family Mystacinidae

Genus *Mystacina*	New Zealand short-tailed bats	
M. *robusta*	New Zealand greater short-tailed bat	Extinct >50 years
M. *tuberculata*	New Zealand lesser short-tailed bat	Vulnerable

Family Molossidae—free-tailed bats

Genus *Chaerephon*	Lesser free-tailed bats	
C. *aloysiisabaudiae*	Duke of Abruzzi's free-tailed bat	No assessment
C. *ansorgei*	Ansorge's free-tailed bat	No assessment
C. *bemmeleni*	Gland-tailed free-tailed bat	No assessment
C. *bivittata*	Spotted free-tailed bat	No assessment
C. *chapini*	Chapin's free-tailed bat	No assessment
C. *gallagheri*	Gallagher's free-tailed bat	No assessment
C. *jobensis*	Northern mastiff bat	No assessment
C. *johorensis*	Dato Meldrum's bat	No assessment
C. *major*	Lappet-eared free-tailed bat	No assessment
C. *nigeriae*	Nigerian free-tailed bat	No assessment
C. *plicata*	Wrinkle-lipped free-tailed bat	No assessment
C. *pumila*	Little free-tailed bat	No assessment
C. *russata*	Russet free-tailed bat	No assessment
Genus *Cheiromeles*		
C. *torquatus*	Hairless bat	Endangered
Genus *Eumops*	Greater mastiff bats	
E. *auripendulus*	Slouch-eared mastiff bat	Stable
E. *bonariensis*	Peters's mastiff bat	Stable
E. *dabbenei*	Big mastiff bat	Stable
E. *glaucinus*	Wagner's mastiff bat	Vulnerable
E. *hansae*	Sanborn's mastiff bat	Stable
E. *maurus*	Guianan mastiff bat	Stable
E. *perotis*	Western mastiff bat	Endangered
E. *underwoodi*	Underwood's mastiff bat	Stable
Genus *Molossops*	Dog-faced bats	
M. *abrasus*	Cinnamon dog-faced bat	Stable
M. *aequatorianus*	Equatorial dog-faced bat	Stable
M. *greenhalli*	Greenhall's dog-faced bat	Stable
M. *mattogrossensis*	South American flat-headed bat	Stable
M. *neglectus*	Rufous dog-faced bat	Stable
M. *planirostris*	Southern dog-faced bat	Stable
M. *temminckii*	Dwarf dog-faced bat	Stable
Genus *Molossus*	Mastiff bats	
M. *ater*	Black mastiff bat	Stable
M. *bondae*	Bonda mastiff bat	Stable
M. *molossus*	Pallas's mastiff bat	Stable
M. *pretiosus*	Miller's mastiff bat	Stable
M. *sinaloae*	Allen's mastiff bat	Stable

Continued

Scientific Name	Common Name	Conservation Status
Family Molossidae—free-tailed bats (*continued*)		
Genus *Mops*	Greater free-tailed bats	
M. brachypterus	Short-winged free-tailed bat	No assessment
M. condylurus	Angola free-tailed bat	No assessment
M. congicus	Medje free-tailed bat	No assessment
M. demonstrator	Mongalla free-tailed bat	No assessment
M. midas	Midas free-tailed bat	No assessment
M. mops	Malayan free-tailed bat	No assessment
M. nanulus	Dwarf free-tailed bat	No assessment
M. niangarae	Niangara free-tailed bat	No assessment
M. niveiventer	White-bellied free-tailed bat	No assessment
M. petersoni	Peterson's free-tailed bat	No assessment
M. sarasinorum	Sulawesi mastiff bat	No assessment
M. spurrelli	Spurrell's free-tailed bat	No assessment
M. thersites	Railer bat	No assessment
M. trevori	Trevor's bat	No assessment
Genus *Mormopterus*	Little mastiff bats	
M. acetabulosus	Natal free-tailed bat	Potentially vulnerable
M. beccarii	Beccari's mastiff bat	No assessment
M. doriae	Sumatran mastiff bat	No assessment
M. jugularis	Peters's wrinkle-lipped bat	No assessment
M. kalinowskii	Kalinowski's bat	Potentially vulnerable
M. minutus	Little mastiff bat	Potentially vulnerable
M. norfolkensis	Eastern little mastiff bat	No assessment
M. petrophilus	Roberts's flat-headed bat	No assessment
M. phrudus	Machu Picchu bat	Stable
M. planiceps	Little flat mastiff bat	No assessment
M. setiger	Peters's flat-headed bat	No assessment
Genus *Myopterus*	African free-tailed bats	
M. daubentonii	Daubenton's free-tailed bat	No assessment
M. whitleyi	Bini free-tailed bat	Potentially vulnerable
Genus *Nyctinomops*	New World free-tailed bats	
N. aurispinosus	Peale's free-tailed bat	Stable
N. femorosaccus	Pocketed free-tailed bat	Stable
N. laticaudatus	Broad-tailed bat	Stable
N. macrotis	Big free-tailed bat	Stable
Genus *Otomops*	Big-eared free-tailed bats	
O. formosus	Java mastiff bat	Potentially vulnerable
O. martiensseni	Large-eared free-tailed bat	No assessment
O. papuensis	Big-eared mastiff bat	No assessment
O. secundus	Mantled mastiff bat	No assessment
O. wroughtoni	Wroughton's free-tailed bat	No assessment
Genus *Promops*	Crested mastiff bats	
P. centralis	Thomas's mastiff bat	Stable
P. nasutus	Brown mastiff bat	Stable
Genus *Tadarida*	Free-tailed bats	
T. aegyptiaca	Egyptian free-tailed bat	No assessment
T. australis	White-striped mastiff bat	No assessment
T. brasiliensis	Brazilian free-tailed bat	Vulnerable
T. espiritosantensis	Espirito Santo free-tailed bat	Stable
T. fulminans	Madagascar large free-tailed bat	No assessment
T. lobata	Kenya big-eared free-tailed bat	Potentially vulnerable
T. teniotis	European free-tailed bat	No assessment
T. ventralis	African giant free-tailed bat	Potentially vulnerable

GENERAL BIBLIOGRAPHY

Allen, G. M. 1939. *Bats*. Cambridge, Mass.: Harvard University Press.

Fenton, M. B. 1983. *Just Bats*. Toronto: University of Toronto Press.

———. 1992. *Bats*. New York: Facts on File.

Findley, J. S. 1993. *Bats: A Community Perspective*. Cambridge: Cambridge University Press.

Fleming, T. H. 1988. *The Short-Tailed Fruit Bat*. Chicago: University of Chicago Press.

Griffin, D. R. 1958. *Listening in the Dark*. New Haven: Yale University Press.

Hill, J. E., and J. D. Smith. 1984. *Bats: A Natural History*. Austin: University of Texas Press.

Kunz, T. H. 1982. *Ecology of Bats*. New York: Plenum Press.

Peterson, R. 1963. *Silently by Night*. London: Longmans.

Ransome, R. 1990. *The Natural History of Hibernating Bats*. London: Christopher Helm.

Richarz, K., and A. Limbrunner. 1993. *The World of Bats*. Neptune City, N.J.: TFH Publications.

Turner, D. C. 1975. *The Vampire Bat: A Field Study in Behavior and Ecology*. Baltimore: Johns Hopkins University Press.

Tuttle, M. D. 1988. *America's Neighborhood Bats*. Austin: University of Texas Press.

Wilson, D. E., and D. M. Reeder, eds. 1993. *Mammal Species of the World*. Washington, D.C.: Smithsonian Institution Press.

Wimsatt, W. A., ed. 1970–77. *The Biology of Bats*. New York: Academic Press, Vols. 1 and 2, 1970; Vol. 3, 1977.

SUBJECT BIBLIOGRAPHY

1. BAT FACTS

How Are Bats Classified?

Bogan, M. A., H. W. Setzer, J. S. Findley, and D. E. Wilson. 1978. Phenetics of *Myotis blythi* in Morocco. Pp. 217–230 in R. J. Olembo, J. B. Castelino, and F. A. Mutere, eds. *Proceedings of the Fourth International Bat Research Conference*. Nairobi, Kenya: National Academy for Advancement of Arts and Sciences.

Koopman, K. F. 1993. Order Chiroptera. Pp. 137–242 in D. E. Wilson and D. M. Reeder, eds. *Mammal Species of the World*. Washington, D.C.: Smithsonian Institution Press.

———. 1994. Chiroptera: Systematics. *Handbook of Zoology*, Vol. 8, Pt. 60. New York: Walter de Gruyter.

How Are Bats Alike?

Baker, R. J., D. C. Carter, and J. K. Jones, Jr. 1976–79. *Biology of Bats of the New World Family Phyllostomatidae*. Lubbock: Special Publications, The Museum, Texas Tech University. Pt. 1, 1976; Pt. 2, 1977; Pt. 3, 1979.

Crerar, L. M., and M. B. Fenton. 1984. Cervical vertebrae in relation to roosting posture in bats. *Journal of Mammalogy* 65 (2): 395–403.

Findley, J. S., E. H. Studier, and D. E. Wilson. 1972. Morphologic properties of bat wings. *Journal of Mammalogy* 53 (3): 429–444.

Findley, J. S., and D. E. Wilson. 1982. Ecological significance of chiropteran morphology. Pp. 243–260 in T. H. Kunz, ed. *Ecology of Bats*. New York: Plenum Press.

Hope, G. M., and K. P. Bhatnagar. 1980. Comparative electroretinography in phyllostomid and vespertilionid bats. Pp. 79–90 in D. E. Wilson and A. L. Gardner, eds. *Proceedings of the Fifth International Bat Research Conference*. Lubbock: Texas Tech Press.

Wilson, D. E. 1985. New mammal records from Sinaloa: *Nyctinomops aurispinosa* and *Onychomys torridus*. *Southwestern Naturalist* 30:303–304.

Wilson, D. E., and A. L. Gardner, eds. 1980. *Proceedings of the Fifth International Bat Research Conference*. Lubbock: Texas Tech Press.

How Do Bats Differ from Other Mammals?

Wilson, D. E. 1989. Bats. Pp. 365–382 in H. Lieth and M. J. A. Werger, eds. *Tropical Rain Forest Ecosystems*. Amsterdam: Elsevier.

What Is Echolocation?

Forbes, B., and E. M. Newhook. 1990. A comparison of the performance of three models of bat detectors. *Journal of Mammalogy* 71 (1): 108–110.

How Do Bats Fly?

Norberg, U. M. 1990. *Vertebrate Flight*. Berlin: Springer-Verlag.

Pennycuick, C. J. 1972. *Animal Flight*. London: Edward Arnold Studies in Biology, No. 33.

Wilson, D. E., and J. Engbring. 1992. The flying foxes *Pteropus samoensis* and *Pteropus tonganus*: Status in Fiji and Samoa. Pp. 74–101 in D. E. Wilson and G. L. Graham, eds. *Pacific Island Flying Foxes: Proceedings of an International Conservation Conference*. U.S. Fish and Wildlife Service Biological Report 90 (23): 74–101.

How Fast Can Bats Fly?

Hayward, B., and R. Davis. 1964. Flight speeds in western bats. *Journal of Mammalogy* 45 (2): 236–241.

Wilson, D. E. 1988. Maintaining bats for captive studies. Pp. 247–264 in T. H. Kunz, ed. *Ecological and Behavioral Methods for the Study of Bats*. Washington, D.C.: Smithsonian Institution Press.

Do Bats Fly in Flocks?

Handley, C. O., Jr., and D. W. Morrison. 1991. Foraging behavior. Pp. 137–140 in C. O. Handley, Jr., D. E. Wilson, and A. L. Gardner. *Demography and Natural History of the Common Fruit Bat, Artibeus jamaicensis, on Barro Colorado Island, Panamá*. Washington, D.C.: Smithsonian Institution Press.

Howell, D. J. 1979. Flock foraging in nectar-feeding bats: Advantages to the bats and to the host plants. *American Naturalist* 114:23–49.

Sazima, I., and M. Sazima. 1977. Solitary and group foraging: Two flower-visiting patterns of the lesser spear-nosed bat *Phyllostomus discolor*. *Biotropica* 9:213–215.

Are All Bats Brown?

Gardner, A. L., and D. E. Wilson. 1970. A melanized subcutaneous covering of the cranial musculature in the phyllostomid bat, *Ectophylla alba*. *Journal of Mammalogy* 52 (4): 854–855.

What Are the Largest and Smallest Bats?

Brosset, A. 1966. *La Biologie des chiroptères*. Paris: Masson et Cie.

How Long Do Bats Live?

Handley, C. O., Jr., D. E. Wilson, and A. L. Gardner. 1991. *Demography and Natural History of the Common Fruit Bat, Artibeus jamaicensis, on Barro Colorado Island, Panamá*. Washington, D.C.: Smithsonian Institution Press.

Wilson, D. E., and E. L. Tyson. 1970. Longevity records for *Artibeus jamaicensis* and *Myotis nigricans*. *Journal of Mammalogy* 51 (1): 203.

Where Do Bats Live?

Findley, J. S., and D. E. Wilson. 1974. Observations on the neotropical disk-winged bat, *Thyroptera tricolor*. *Journal of Mammalogy* 55 (3): 562–571.

———. 1984. Are bats rare in tropical Africa? *Biotropica* 15:299–303.

Gardner, A. L., R. K. LaVal, and D. E. Wilson. 1970. The distributional status of some Costa Rican bats. *Journal of Mammalogy* 51 (4): 712–729.

Hensley, D. 1993. Motorola supports bat house research project. *Bats* 11 (4): 16.

Mok, W. Y., L. A. Lacey, R. C. Luizao, and D. E. Wilson. 1982. Lista atualizada de quirópteros da Amazônia Brasileira. *Acta Amazonica* 12:817–823.

Studier, E. H., and D. E. Wilson. 1983. Natural urine concentrations and composition in neotropical bats. *Comparative Biochemistry and Physiology* 75A:509–515.

Studier, E. H., S. J. Wisniewski, A. T. Feldman, R. W. Dapson, B. C. Boyd, and D. E. Wilson. 1983. Kidney structure in neotropical bats. *Journal of Mammalogy* 64 (3): 445–452.

Tuttle, M. D., and D. L. Hensley. 1993. *The Bat House Builder's Handbook*. Austin, Tex.: Bat Conservation International.

Tuttle, M. D., and D. A. R. Taylor. 1994. *Bats and Mines*. Austin, Tex.: Bat Conservation International Resource Publication, No. 3: 1–41.

Wilson, D. E. 1970. An unusual roost of *Artibeus cinereus watsoni*. *Journal of Mammalogy* 51 (1): 204–205.

———. 1973. Bat faunas: A trophic comparison. *Systematic Zoology* 22 (1): 14–29.

Wilson, D. E., and J. S. Findley. 1972. Randomness in bat homing. *American Naturalist* 106:418–424.

Wilson, D. E., R. A. Medellín, D. V. Lanning, and H. T. Arita. 1985. Los Murciélagos del noreste de México, con una lista de especies. *Acta Zoologica Mexicana* 8:1–26.

How Do Bats Reproduce?

Altenbach, J. S., K. N. Geluso, and D. E. Wilson. 1976. Bat mortality: Pesticide poisoning and migratory stress. *Science* 194:184–186.

Fleming, T. H., E. T. Hooper, and D. E. Wilson. 1972. Three Central American bat communities: Structure, reproductive cycles, and movement patterns. *Ecology* 53 (4): 555–569.

Mares, M. A., and D. E. Wilson. 1971. Bat reproduction during the Costa Rican dry season. *BioScience* 21:471–477.

Racy, P. A. 1982. Ecology of bat reproduction. Pp. 57–104 in T. H. Kunz, ed. *Ecology of Bats*. New York: Plenum Press.

Wilson, D. E. 1973. Reproduction in neotropical bats. *Periodicum Biologorum* 75:215–217.

———. 1983. *Myotis nigricans* (Murciélago pardo, black myotis). Pp. 477–478 in D. H. Janzen, ed. *Costa Rican Natural History*. Chicago: University of Chicago Press.

Wilson, D. E., and J. S. Findley. 1970. Reproductive cycle of a neotropical insectivorous bat, *Myotis nigricans*. *Nature* 225:1155.

———. 1971. Spermatogenesis in some neotropical species of *Myotis*. *Journal of Mammalogy* 52 (2): 420–426.

Wilson, D. E., C. O. Handley, Jr., and A. L. Gardner. 1991. Reproduction on Barro Colorado Island. Pp. 43–52 in C. O. Handley, Jr., D. E. Wilson, and A. L. Gardner. *Demography and Natural History of the Common Fruit Bat, Artibeus jamaicensis, on Barro Colorado Island, Panamá*. Washington, D.C.: Smithsonian Institution Press.

Wilson, D. E., and R. K. LaVal. 1974. *Myotis nigricans*. *Mammalian Species* 39:1–4.

How and Where Do Mother Bats Give Birth?

Wilson, D. E. 1971. Ecology of *Myotis nigricans* (Mammalia: Chiroptera) on Barro Colorado Island, Panama Canal Zone. *Journal of Zoology* 163 (1): 1–13.

How Long Do Mother Bats Suckle Their Young?

Wilson, D. E. 1979. Reproductive patterns. Pp. 317–378 in R. J. Baker, J. K. Jones, Jr., and D. C. Carter, eds. *Biology of Bats of the New World Family Phyllostomatidae*. Special Publications, The Museum, Texas Tech University. Pt. 3, No. 16.

How Fast Do Bats Grow?

Adams, R. A., and S. C. Pedersen. 1994. Wings on their fingers. *Natural History* 103 (1): 48–54.

What Do Bats Eat?

Fascione, N., T. Marceron, and M. B. Fenton. 1991. Evidence for mosquito consumption in M. *lucifugus*. *Bat Research News* 32 (1): 2–3.

Ferrell, C. S., and D. E. Wilson. 1991. *Platyrrhinus helleri*. *Mammalian Species* 373:1–5.

Medellín, R. A., D. Navarro-L., and D. E. Wilson. 1985. *Micronycteris brachyotis*. *Mammalian Species* 251:1–4.

Mizutani, H., D. A. McFarlane, and Y. Kabaya. 1992. Nitrogen and carbon isotope study of bat guano core from Eagle Creek Cave, Arizona, U.S.A. *Mass Spectroscopy* 40 (1): 57–65.

Rydell, J. 1989. Food habits of northern (*Eptesicus nilssoni*) and brown long-eared (*Plecotus auritus*) bats in Sweden. *Holarctic Ecology* 12:16–20.

Studier, E. H., B. C. Boyd, A. T. Feldman, R. W. Dapson, and D. E. Wilson. 1983. Renal function in the neotropical bat *Artibeus jamaicensis*. *Comparative Biochemistry and Physiology* 74A:199–209.

Whitaker, J. O., Jr. 1993. Bats, beetles, and bugs. *Bats* 11 (1): 23.

Wilson, D. E. 1990. Mammals of La Selva, Costa Rica. Pp. 273–286 in A. H. Gentry, ed. *Four Neotropical Forests*. New Haven: Yale University Press.

———. 1991. Mammals of the Tres Marias Islands. Pp. 214–250 in T. A. Griffiths and D. Klingener, eds. *Contributions to Mammalogy in Honor of Karl F. Koopman*. *Bulletin of the American Museum of Natural History*, No. 206.

Wilson, D. E., and I. Gamarra de Fox. 1991. El Murciélago *Macrophyllum macrophyllum* en Paraguay. *Boletín de Museo Nacional de Historia Natural del Paraguay* 10:33–35.

Zortéa, M., and S. L. Mendes. 1993. Folivory in the big fruit-eating bat, *Artibeus lituratus* (Chiroptera: Phyllostomidae) in eastern Brazil. *Journal of Tropical Ecology* 9:117–120.

How Do Bats Find Food?

Tuttle, M. D. 1982. The amazing frog-eating bat. *National Geographic* 161 (1): 78–91.

Wilson, D. E. 1971. Food habits of *Micronycteris hirsuta* (Chiroptera: Phyllostomidae). *Mammalia* 35 (1): 107–110.

How Smart Are Bats?

Baron, G., and P. Jolicoeur. 1980. Brain structure in Chiroptera: Some multivariate trends. *Evolution* 34:386–393.

Eisenberg, J. H., and D. E. Wilson. 1978. Relative brain size and feeding strategies in the Chiroptera. *Evolution* 32:740–751.

What Do Bats Do in Winter?

Studier, E. H., and D. E. Wilson. 1970. Thermoregulation in some neotropical bats. *Comparative Biochemistry and Physiology* 34:252–262.

———. 1979. Effects of captivity on thermoregulation and metabolism in *Artibeus jamaicensis* (Chiroptera: Phyllostomatidae). *Comparative Biochemistry and Physiology* 62B:347–350.

Why Do Some Bats Migrate?

Geluso, K. N., J. S. Altenbach, and D. E. Wilson. 1981. Organochlorine residues in young Mexican free-tailed bats from several roosts. *American Midland Naturalist* 105:249–257.

Timm, R. M., D. E. Wilson, B. L. Clawson, R. K. LaVal, and C. M. Vaughan. 1989. Mammals of the Braulio Carrillo–La Selva Complex, Costa Rica. *North American Fauna*, No. 75.

Wilson, D. E. 1980. Murciélagos migratorios en el continente Americano. Organización de los Estados Americanos, Reunión técnica sobre conservación de animales migratorios des hemisferio occidental y sus ecosistemas. SG/Ser.P/III.3: 62–63.

Do Bats Have Enemies?

Johns, A. D., R. H. Pine, and D. E. Wilson. 1985. Rainforest bats: An uncertain future. *Bat News* 5:4–5.

Wilson, D. E. 1990. Pacific flying foxes surveyed. *Endangered Species Technical Bulletin* 15:4.

2. BAT EVOLUTION AND DIVERSITY

When Did Bats Evolve?

Jepsen, G. L. 1980. Bat origins and evolution. Pp. 1–64 in W. A. Wimsatt, ed. *Biology of Bats*. New York: Academic Press, Vol. 1.

Simmons, N. 1994. The case for chiropteran monophyly. *American Museum Novitates* 3103:1–54.

Smith, J. D. 1980. Chiropteran phylogenetics: Introduction. Pp. 233–244 in D. E. Wilson and A. L. Gardner, eds. *Proceedings of the Fifth International Bat Research Conference*. Lubbock: Texas Tech Press.

Van Valen, L. 1979. The evolution of bats. *Evolutionary Theory* 4:103–121.

Where Are Fossil Bats Found?

Ray, C. E., and D. E. Wilson. 1979. Evidence for *Macrotus californicus* from Terlingua, Texas. *Occasional Papers, The Museum, Texas Tech University* 57:1–10.

How Many Species of Bats Are There?

Ascorra, C. F., D. E. Wilson, and M. Romo. 1991. Lista anotada de los quirópteros del Parque Nacional Manu, Perú. *Publicaciones del Museo de Historia Natural, Universidad Nacional Mayor de San Marcos, Ser. A Zoologia* 42:1–14.

Engstrom, M. D., T. E. Lee, and D. E. Wilson. 1987. *Bauerus dubiaquercus*. *Mammalian Species* 282:1–3.

Engstrom, M. D., and D. E. Wilson. 1981. Systematics of *Antrozous dubiaquercus* (Chiroptera: Vespertilionidae), with comments on the status of Bauerus Van Gelder. *Annals of Carnegie Museum* 50:371–383.

Greenbaum, I. F., R. J. Baker, and D. E. Wilson. 1975. Evolutionary implications of the karyotypes of the Stenodermine Genera *Ardops, Ariteus, Phyllops,* and *Ectophylla*. *Bulletin of Southern California Academy of Science* 74:156–159.

Hill, J. E. 1974. A new family, genus, and species of bat (Mammalia: Chiroptera) from Thailand. *Bulletin of the British Museum (Natural History), Zoology* 27:301–336.

Lim, B. K., and D. E. Wilson. 1993. Taxonomic status of *Artibeus amplus* (Chiroptera: Phyllosto-midae) in northern South America. *Journal of Mammalogy* 74 (3): 763–768.

Wilson, D. E., and J. A. Salazar. 1990. Los Murciélagos de la reserva de la biósfera "Estación Bi-ológica Beni." *Ecologia en Bolivia* 13:47–56.

What Characterizes the Major Groups of Bats?

Lassieur, S., and D. E . Wilson. 1989. *Lonchorhina aurita*. *Mammalian Species* 347:1–4.

Wilson, D. E., and G. Graham, eds. 1992. *Pacific Island Flying Foxes: Proceedings of an International Conservation Conference*. U.S. Fish and Wildlife Service Biological Report 90 (23): 1–176.

Flying Foxes

DeFrees, S. L., and D. E. Wilson. 1988. *Eidolon helvum*. *Mammalian Species* 312:1–5.

Nowak, R. M. 1991. *Walker's Mammals of the World*. 5th ed. Baltimore: Johns Hopkins Univer-sity Press, Vol. 1.

Blossom Bats

Strahan, R. 1995. *Complete Book of Australian Mammals*. Washington, D.C.: Smithsonian Insti-tution Press.

Mouse-Tailed Bats

Lekagul, B., and J. A. McNeely. 1977. *Mammals of Thailand*. Bangkok: Association for the Con-servation of Wildlife, Sahakarnbhat.

Qumsiyeh, M. B., and J. K. Jones, Jr. 1986. *Rhinopoma hardwickii* and *Rhinopoma muscatellum*. *Mammalian Species* 263:1–5.

Hog-Nosed Bats

Hill, J. E., and S. E. Smith. 1981. *Craseonycteris thonglongyai*. *Mammalian Species* 160:1–4.

White Bats

Ceballos-G., G., and R. A. Medellín. 1988. *Diclidurus albus*. *Mammalian Species* 325:1–4.

Slit-Faced Bats

Rosevear, D. R. 1965. *The Bats of West Africa*. London: British Museum (Natural History).

Old World False Vampire Bats

Hudson, W. S., and D. E. Wilson. 1986. *Macroderma gigas*. *Mammalian Species* 260:1–4.

Horseshoe Bats

Kingdon, J. 1974. *East African Mammals: An Atlas of Evolution in Africa*, Vol. 2(A). *Insectivores and Bats*. London: Academic Press.

Roundleaf Bats

Hill, J. E., and J. D. Smith. 1984. *Bats: A Natural History*. Austin: University of Texas Press.

Bulldog Bats

Hood, C. S., and J. K. Jones, Jr. 1984. *Noctilio leporinus*. *Mammalian Species* 216:1–7.

Hood, C. S., and J. Pitocchelli. 1983. *Noctilio albiventris*. *Mammalian Species* 197:1–5.

Naked-Backed Bats

Herd, R. M. 1983. *Pteronotus parnellii*. *Mammalian Species* 209:1–5.

Spectral Vampire Bats

Ascorra, C. F., D. E. Wilson, and A. L. Gardner. 1991. Geographic distribution of *Micronycteris schmidtorum* Sanborn (Chiroptera: Phyllostomidae). *Proceedings of the Biological Society of Washington* 104:351–355.

Navarro, D., and D. E. Wilson. 1982. *Vampyrum spectrum*. *Mammalian Species* 184: 1–4.

Nectar-Feeding Bats

Arita, H. T., and D. E. Wilson. 1987. Long-nosed bats and agaves: The tequila connection. *Bats* 5 (4): 3–5.

———. 1990. Operación Tequila: Los Murciélagos narigudos y el agave. *México Desconocido* 13:25–31.

Flower Bats

Nellis, D. W., and C. P. Ehle. 1977. Observations on the behavior of *Brachyphylla cavernarum* (Chiroptera) in Virgin Islands. *Mammalia* 41 (4): 403–409.

Long-Tongued Bats

Hill, J. E., and J. D. Smith. 1984. *Bats: A Natural History*. Austin: University of Texas Press.

Short-Tailed Fruit Bats

Fleming, T. H. 1988. *The Short-Tailed Fruit Bat*. Chicago: University of Chicago Press.

Wilson, D. E. 1990. Review of *The Short-Tailed Fruit Bat: A Study in Plant-Animal Interactions* (by Theodore H. Fleming). *Journal of Wildlife Management* 54:369–370.

American Epauletted Bats

Nowak, R. M. 1991. *Walker's Mammals of the World*. 5th ed. Baltimore: Johns Hopkins University Press, Vol. 1.

Tent-Making Bats

Baker, R. J., and C. L. Clark. 1987. *Uroderma bilobatum*. *Mammalian Species* 279:1–4.

Barbour, T. 1932. A peculiar roosting habit of bats. *Quarterly Review of Biology* 7:307–312.

Foster, M. F., and R. M. Timm. 1976. Tent-making by *Artibeus jamaicensis* (Chiroptera: Phyllostomatidae) with comments on plants used by bats for tents. *Biotropica* 8:265–269.

Lewis, S. E., and D. E. Wilson. 1987. *Vampyressa pusilla*. *Mammalian Species* 292:1–5.

Vampire Bats

Altenbach, J. S. 1979. *Locomotor Morphology of the Vampire Bat, Desmodus rotundus*. Special Publication 6, American Society of Mammalogists.

Greenhall, A. M., G. Joermann, U. Schmidt, and M. R. Seidel. 1983. *Desmodus rotundus*. *Mammalian Species* 202:1–6.

Turner, D. C. 1975. *The Vampire Bat: A Field Study in Behavior and Ecology*. Baltimore: Johns Hopkins University Press.

Wilson, D. E., 1980. Locomotor morphology of the vampire bat, *Desmodus rotundus*. (Review.) *Association of Systematics Collections Newsletter* 8:14–15.

Funnel-Eared Bats

Hill, J. E., and J. D. Smith. 1984. *Bats: A Natural History*. Austin: University of Texas Press.

Thumbless Bats

LaVal, R. K. 1977. Notes on some Costa Rican bats. *Brenesia* 10–11: 77–83.

Sucker-Footed Bats

Hill, J. E., and J. D. Smith. 1984. *Bats: A Natural History*. Austin: University of Texas Press.

Disk-Winged Bats

Wilson, D. E. 1976. The subspecies of *Thyroptera discifera* (Lichtenstein and Peters). *Proceedings of the Biological Society of Washington* 89:305–312.

———. *Thyroptera discifera*. *Mammalian Species* 104:1–3.

Wilson, D. E., and J. S. Findley. 1977. *Thyroptera tricolor*. *Mammalian Species* 71:1–3.

Painted Bats

Nowak, R. M. 1991. *Walker's Mammals of the World*, 5th ed. Baltimore: Johns Hopkins University Press, Vol. 1.

Little Brown Bats

Fenton, M. B., and R. M. R. Barclay. 1980. *Myotis lucifugus*. *Mammalian Species* 142:1–8.

Tube-Nosed Insectivorous Bats

Nowak, R. M. 1991. *Walker's Mammals of the World*, 5th ed. Baltimore: Johns Hopkins University Press, Vol. 1.

Long-Fingered Bats

Van Der Merwe, M. 1978. Postnatal development and mother-infant relationships in the Natal clinging bat *Miniopterus schreibersi natalensis* (A. Smith, 1834). *Proceedings of the Fourth International Bat Research Conference*. Nairobi, Kenya: National Academy for Advancement of Arts and Sciences. 1:309–322.

Short-Tailed Bats

Fenton, M. B. 1992. *Bats*. New York: Facts on File.

Free-Tailed Bats

Altenbach, J. S., K. N. Geluso, and D. E. Wilson. 1979. Population size of *Tadarida brasiliensis* at Carlsbad Caverns in 1973. Pp. 341–348 in H. H. Genoways and R. J. Baker, eds. *Biological Investigations in the Guadalupe Mountains National Park, Texas*. U.S. National Park Service Proceedings and Transactions Series, No. 4.

Ascorra, C. F., D. E. Wilson, and C. O. Handley, Jr. 1991. Geographic distribution of *Molossops neglectus* Williams and Genoways (Chiroptera: Molossidae). *Journal of Mammalogy* 72 (4): 828–830.

Wilson, D. E., K. N. Geluso, and J. S. Altenbach. 1978. The ontogeny of fat deposition in *Tadarida brasiliensis*. R. J. Olembo, J. B. Castelino, and F. A. Mutere, eds. *Proceedings of the Fourth International Bat Research Conference*. Nairobi, Kenya: National Academy for Advancement of Arts and Sciences. 1:15–20.

3. BATS AND HUMANS

How Can I Attract Bats?

Tuttle, M. D., and D. L. Hensley. 1993. *The Bat House Builder's Handbook*. Austin, Tex.: Bat Conservation International.

What Are Some Myths and Legends about Bats?

McCracken, G. F. 1993. Folklore and the origin of bats. *Bats* 11 (4): 11–13.

What Good Are Bats?

Ascorra, C. F., and D. E. Wilson. 1992. Bat frugivory and seed dispersal in the Amazon, Loreto, Perú. *Publicaciones del Museo de Historia Natural, Universidad Nacional Mayor de San Marcos, Ser. A Zoologia* 43:1–6.

How Do Scientists Catch Bats?

Kunz, T. H., ed. 1988. *Ecological and Behavioral Methods for the Study of Bats*. Washington, D.C.: Smithsonian Institution Press.

TAXONOMIC INDEX

Taxa are indexed by their core common names, with the scientific equivalent following in parentheses. For example, "Jamaican fruit bats" are listed as "fruit bats, Jamaican." Page-number citations in italics refer to *figures* on those pages. For a comprehensive listing of all known bat species, see the appendix.

SUBJECT INDEX

Page-number citations in italics refer to *figures*, and those in boldface refer to **tables** on those pages. Bat names are found in the taxonomic index.

anatomy, 3–14, *4–13*
anticoagulant in saliva, 59–60, 101, 123
attitudes toward bats, 118, *118*, 119
attracting bat colonies, 116–117, *117*

banding, 30
Bat Conservation International, 128, 130
biologist, studying bats, 124–125, *125*
birds versus bats, 14–15, *14*
"blind as a bat" myth, 9
blood feeders, 58–60, 101–102
blossom (flower) feeders, 50–51, *52*, 86, *86*
body shape, 3, *4*, *5*
bony structure, 3, *4*, 5–6, *5*
    adaptations for flying, 14–15, *14*
    fossils, 79–81, *81*
brain, 6, 63–64
breeding. *See* reproduction
buildings
    bat removal from, 115
    roosting in, 115

cannibalism, 57
careers in bat biology, 124–125, *125*
carnivores, 56–58, *57*, 93–94
catching bats, 125–126, *126*, *127*
caves
    bat gates for, 34, *34*
    fossils in, 81
    funnel traps for, 126, *127*

roosting in, 32–34, *33*, 109–111
classification, 2, **3.** *See also* the appendix *and* the
        taxonomic index
    knowledge deficits in, 126, 128
    species, 81–82
colonies, 32–34, *33*
    in attic, 115–116
    attracting, 116–117, *117*
    nursery, 107
color, 4, 6, 23, *24*, 25–26, 106
communication, 105. *See also* echolocation;
        vocalization
conservation
    information sources, 128, 130–131
    reasons for, 120–121, *120–123*, 123
control methods, 116
countershading, 23, *23*
courtship, 40–41, *41*
crests, of skull, 6
cultural attitudes toward bats, *118*, 119

dentition. *See* teeth
development of young bats, 45–46, *45*
diet. *See* food and eating

ears, 11–12, *11*
    in echolocation, 17–18
echolocation, 15–18, *16*, 28
    ear function in, 11, 17–18
    in fish catching, 58, 59
    in insect catching, 51–52, 62–63, *62*
e-mail discussion group (Batline), 131
employment as bat biologist, 124–125, *125*
enemies, 70–72, *71*, *72*

166

environment, interactions with, 128, *129*
epaulets, 23
evolution, 79–81
eyes, 9–10, *10*

fish eaters, 58, 59, 92
fleas, 114–115
flocking, 21
flower feeders, 50–51, *52*, 86, *86*
flying, 18–20, *19, 20. See also* wings
   enemy predation during, 70, *71*
   evolution of, 80
   in flocks, 21
   into people's hair, 113–114
   soaring in, 18–19, *20*
   speed of, 20
   of young bats, 45–46
foliage. *See* leaves
folk tales, 9, 113–114, 119
food and eating, 47–62
   aerial insects, 51–54, *53*, 61–63
   blood, 58–60, 101–102
   bulldog bats, 91
   false vampire bats, 89
   fish, 58, *59*, 92
   flowers, 50–51, *52*, 86, *86*
   flying foxes, 84–85, *84*
   foliage insects, 54–56, *54–56*, 62
   frogs, 57–58, *57*, 63
   fruit, 48–50, *48–49, 50*, 61, *61*
   for hibernation preparation, 66, *67*
   jaw structure and, 6–8, *7, 8*
   roundleaf bats, 91
   search for, 61–63, *61, 62*
   short-tailed fruit bats, 98–99
   vampire bats, 101–102
   vertebrates, 56–58, *57*
form, 3, *4, 5*
fossil bats, 79–81, *81*
frog eaters, 57–58, *57*, 63
fruit eating, 48–50, *48–49, 50*, 61, *61*
funnel traps, 126, *127*
fur, 4–5

gates for protecting bat colonies, 34, *34*
geographic distribution
   of extant bats, 30–32
   of fossil bats, 81, *81*
glands on shoulders, 99–100
gliding, 37, 80
growth of young bats, 45–46, *45*
guano deposits
   economic value of, 120, *121*

geochronology of, 53–54
   human illness from, 73

habitat, 30–32
hands, 3, *4*
health problems, 72–73, 114–115, 117
hearing, 11–12
   in echolocation, 17–18
hibernation, 21–22, 64, 65
   after migration, 68–69
   physiology of, 67–68, *67*
   reasons for, 64, *65–67, 66*
   reproduction timing and, 38
histoplasmosis, 73, 114
homing, 32
houses built for bat colonies, 116–117, *117*

illness, 72-73, 114–115, 117
information sources, 128, 130–131
insect eaters
   aerial, 51–54, *53*, 61–63
   foliage, 54–56, *54–56*, 62
intelligence, 63–64
Internet discussion group (Batline), 131

jaws, 6–8, *7, 8*
Journal of Mammalogy, 124–125
jumping, 59

larynx in echolocation, 15–16
leaves
   insect gleaning from, 54–56, *54–56*, 62
   roosting in, 100–101, 104, *105*
   as shelter, 35–36, *35*
legends, 9, 113–114, 119
life span, 30, *31*
litter size, 43–44, *44*
livestock, vampire bat feeding on, 101–102
locomotion. *See also* flying
   gliding, 37, 80
   on ground, 46
longevity, 30, *31*

mammalogy, study of, 124–125
mating, 38, 40–41, *41*
membranes, wing (uropatagia), 4, 6
migration, 64, 65, 68–69, *69*
mites, 114–115
mosquitoes, consumption of, 53
myths, 9, 113–114, 119

neck, 12
nectar feeders, 50–51, *52*

nets for bat catching, 125–126, *126*
newborns, 42–43, *42, 43*
    number of, 43–44, *44*
    weight of, 45, *45*
nursing, 9, 43–45, *44*
    mother bats' recognition of their own young,
        46–47, *47*
nutrition. *See* food and eating

organizations for bat information, 128, 130–131

pain sensation, 69–70
parasites, 114–115
pesticides, accumulation in fat, 68, 110
pigmentation. *See* color
pollen feeders, 50–51, *52*
pollination, 95, *95, 96, 97*, 121, *122, 123*
polyestry, 39–40
predation
    by bats, 56–58, *57*
    on bats, 70–72, *71, 72*
protection
    bat gate in, 34, *34*
    reasons for, 120–121, *120–123, 123*

rabies, 73, 102, 114
repellents for use against bats, 116
reproduction, 38–41, *39, 41*
    birth, 42, *42, 43*
    brown bats, 107
    bulldog bats, 91
    horseshoe bats, 90
    litter size, 43–44, *44*
    long-fingered bats, 108
    roundleaf bats, 91
    short-tailed bats, 109
roosting, 21–22, 25, *26*, 32
    artificial structures designed for, 116–117, *117*
    birth during, 42
    blossom feeders, 86, *86*
    brown bats, 106–107
    in buildings, 115
    bulldog bats, 92
    disk-winged bats, 104, *105*
    enemy predation during, 70
    false vampire bats, 89
    flying foxes, 84, *84*
    free-tailed bats, 109–110
    in hibernation. *See* hibernation
    horseshoe bats, 90
    in leaves, 100–101, 104, *105*
    mouse-tailed bats, *87*, 88
    places for, 35–36, *35, 36*. *See also* caves

roundleaf bats, 90
slit-faced bats, 89
upside down, 36–37, *37*
vampire bats, 101

saliva, anticoagulant in, 59–60, 101, 123
seeds, dispersal of, 7, 49–50, *50*, 121, *122, 123*
senses
    hearing, 11–12, 17–18
    sight, 9–10, *10*
    touch (pain sensation), 69–70
sex. *See* reproduction
shelter, 35–36, *35, 36*. *See also* caves
sickness, 72–73
size, 29–30
skeleton. *See* bony structure
skin, 4–5, 6
    of wings, 12
skull, 5–6
sleep, 21–22
social systems, 32–33
sound in echolocation. *See* echolocation

teeth, 7, 8–9, *9*
    of blossom feeders, 86
    evolution of, 80
    of vampire bats, 59, 101
temperature regulation, 21–22
    in hibernation, 67–68
ticks, 114–115
trapping bats, 125–126, *126, 127*
trees as shelter, 35, *35*
Tuttle trap, 126, *127*

ultrasonic sounds. *See* echolocation
uropatagia, 4, 6

vision, 9–10, *10*
vocalization, 16, 28, *29*. *See also* echolocation
voice box in echolocation, 15–16

walking, 58
weight of young bats, 45–46, *45*
wings, 3, *4*, 4–5, 6, 12, *13*, 14
    aerodynamics of, 18, *19*
    color of, 25
    comparison with bird wings, 14–15, *14*
    evolution of, 80
    size of, 29
    uniqueness among mammals, 15
winter
    hibernation during. *See* hibernation
    migration during, 64, 65, 68–69, 69